The Patrick Moore Practical Astronomy Series

More information about this series at http://www.springer.com/series/3192

Imaging Sunlight Using a Digital Spectroheliograph

Ken M. Harrison

 Springer

Ken M. Harrison
St Leonards, Vic, Australia

ISSN 1431-9756   ISSN 2197-6562   (electronic)
The Patrick Moore Practical Astronomy Series
ISBN 978-3-319-24872-1   ISBN 978-3-319-24874-5   (eBook)
DOI 10.1007/978-3-319-24874-5

Library of Congress Control Number: 2016937360

Printed on acid-free paper

This Springer imprint is published by Springer Nature
The registered company is Springer International Publishing AG Switzerland

Dedicated to Fred Nall Veio
He kept the dream alive.
In Memory of
Walter Koch (Swisswalter)
5 Jan 1954—27 July 2015
"Only stardust in the wind"

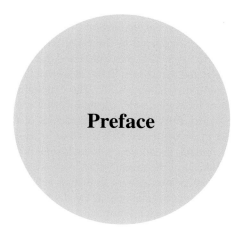

Preface

Amateur astronomers have always been active solar observers. The recent advent of specialized narrowband solar telescopes such as the instruments from Coronado and Lunt has opened up the opportunity of observing and imaging the Sun in Hα and CaK wavelengths. Some of the images produced by solar imagers are of excellent quality and can stand comparison with those taken by professional observatories only a few years ago.

Moving from the traditional "white light" observation of sunspots and surface granulation to this exciting frontier of filaments and prominences, Ellerman bombs and flares unfortunately are still an expensive route to follow. The basic "cost of entry" to the Hα arena is in the $1000 plus range.

In the opening chapter of this book, we will follow the early work of the nineteenth-century amateurs and the introduction of photography to solar imaging before discussing the current designs of safe filters available to the amateur.

In Chap. 3, we review the nature of the Sun, our nearest star, and the varied solar features of interest which are regularly targeted by the imager. We will also provide comprehensive comparisons of the commercial filters and their benefits. Moving from the arena of solar filters, we consider the solar spectrum and the scientific benefits which can be obtained by imaging and recording detail in the spectrum.

This introduction to the solar spectrum then leads on to a parallel and, in my opinion, a more versatile and cost-effective solution to the traditional solar filters— the spectroheliograph (SHG). This instrument, invented in 1890 and further refined for visual observation in the 1920s, has long been the province of the hard-core DIY enthusiasts with avid talk about Anderson prisms and nodding mirrors and secondary optics like heliostats, siderostats, and cœlostats and other such large instruments taking up the whole backyard.

Fred Nall Veio has been championing the spectrohelioscope (SHS) for the past 50 years and has assisted many amateurs to build their own instrument. Without his continued dedication, this instrument and the opportunities it affords the amateur community could well have been lost to posterity.

A significant breakthrough occurred in the early 2000s when the early digital cameras were applied to the SHG. This innovation and the development of supporting software have reinvigorated the interest in the instrument and led the way to even more exciting designs and outstanding results. The concluding chapters cover the design and construction of the digital SHG illustrated with examples of instruments being used around the world, finishing with some details of the spectrohelioscope which will be of interest to the dedicated DIY amateur looking for a challenge.

As we will see, the images produced by today's SHGs are not yet achieving results at the quality standards of the cutting-edge narrowband filters, but it's just a matter of time before the software matures to further refine the images and show the full potential, at any chosen wavelength, of the digital SHG.

This is the story of the digital SHG, being written and continuously updated by amateurs around the globe—following in Fred's footsteps.

St. Leonards, Vic, Australia   Ken M. Harrison

Acknowledgments

Many amateurs around the world have contributed their work to make this book as comprehensive as it is. I thank all those who have allowed me permission to use their illustrations and other materials.

I must mention specifically André Rondi, Philippe Rousselle, Daniel Defourneau, Jean-Jacques Poupeau, and Wah-Heung Yeun. This dedicated group of amateurs has led the way in the ongoing development of the digital SHG, and without their ongoing efforts, we would not be achieving the results we see today.

The active imagers on the Solar Chat forum have been especially supportive in allowing the use of their solar images.

I would also like express my deep appreciation to Fred Nall Veio for keeping the dream alive.

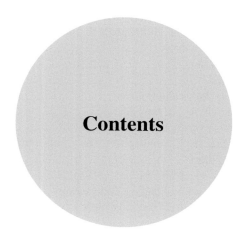

Contents

Chapter 1

Brief History of Solar Observations

1.1 Introduction

Since the earliest times on record solar observing has been practiced throughout the world as a means of understanding the seasons; megalithic monuments like Stonehenge and the solar alignment of the access tunnels to the burial barrows (the passage tombs of Maes Howe and Newgrange immediately come to mind) stand as evidence of the importance of the Sun to our ancestors. Our lifestyle has been determined by the daily rising and setting of the Sun. The solar phenomena of total eclipses have influenced governments and counties worldwide—and unfortunately for some early astronomers who got their calculations wrong also cost them their heads. During the reign of Tchong-Kang in China, in 2155 BC two unfortunate mathematicians Ho and Hi failed to accurately predict a solar eclipse and were put to death for their negligence.

The first measurement of the size of the Earth was achieved by the early Greek astronomer Eratosthenes in 290 BC just by observing the different altitude of the sun from different locations. The Chinese records show that naked eye sunspot groups have been observed for thousands of years. In the Western world, meanwhile, the main item of note was the publication of Copernicus' famous De Revolutionibus in the mid 1500s, stating that "in the center of everything the sun must reside; in the most beautiful temple created by God, there is the place that which awaits him where he can give light to all the planets".

For thousands of years, these naked eye observations and the deductions made from them satisfied the needs of the time, but the invention of the telescope from 1608 to 1610 was to change our perception of the Sun forever.

© Springer International Publishing Switzerland 2016
K.M. Harrison, *Imaging Sunlight Using a Digital Spectroheliograph*,
The Patrick Moore Practical Astronomy Series, DOI 10.1007/978-3-319-24874-5_1

1.1.1 Early Telescopic Observations

White light telescopic observations of the Sun have been conducted since the time of Galileo. Galileo Galilei (1564–1642) and his early contemporary observers were already aware of sunspots. Sometimes through smoke or haze they could be observed by the naked eye; the Chinese had been recording such sunspots for almost 2000 years but had no idea of their real origin, most of the observations being attributed to transits of Venus or Mercury or some other body in the solar system. This probably sounds strange with the benefit of hindsight based on our understanding of today, when scientists know that solar transit (where the disk of a planet appears to move across the surface of the Sun) of either of the inner planets only lasts a few hours whereas naked eye sunspots were visible for a few days.

Even the earliest telescopes were able to reveal that the sunspots were actually visible dark markings on the solar surface which rotated with the Sun. Galileo wrote in 1612:

> "Having made repeated observations I am at last convinced that the spots are objects close to the surface of the solar globe, where they are continually being produced and then dissolved, some quickly and some slowly; also that they are carried round the Sun by its rotation, which is completed in a period of about one lunar month. This is an occurrence of the first importance in itself, and still greater in its implications".

At a time when the Catholic Church would have everyone believe that the Earth was the center of all things and that everything revolved around it—this was pretty dangerous stuff. When Galileo reported he had observed the moons of Jupiter in 1610, then later the phases of Venus in 1613—"Venus imitates the aspects of the moon", the evidence for the old Ptolemaic earth scheme of things was becoming quickly untenable.

For the next few years the solar disk was regularly explored by telescope and it was quickly established that the outer edge of the solar disk appeared darker than the center of the disk. Sunspots were being recorded in sufficient detail to show the darker umbra and lighter penumbra regions and Galileo even recognized the slightly brighter surface areas near sunspot groups, giving them the name *faculae*, meaning "little torch" in Latin. Galileo also measured the brightness of the sunspots. He determined that they were not as dark as the moon shadows but actually as bright as a full moon—only appearing dark on the solar disk due to the extreme brilliance of the surrounding surface. The rotational period of the Sun, and even the approximate inclination of the axis were quickly established, all within a few short years. Such was the power of the new instrument to clearly see what had previously been undistinguishable, and to allow for new interpretations based on observation.

1.1.2 Solar Eclipses

The 5° inclination of the moon's orbit, combined with the Earth's movement around the Sun, means that the moon can sometimes obstruct the view of the disk of the Sun from the perspective of observers on the Earth. When this happens we have a solar eclipse. There are at least two eclipses visible from some part of the Earth every year. Depending on the moon's distance and the alignment of the lunar orbital plane, the eclipse can be partial, where the moon only covers a small amount of the Sun, or total. When the moon is further from the Earth, at lunar apogee, the apparent diameter of the moon is less than the Sun and gives rise to annular eclipses where a broad ring of the Sun is left visible. At other times when the moon is closer to the Earth, or lunar perigee, the moon can appear to be larger than the solar disk and gives us the spectacular total solar eclipse. A total eclipse can have a duration of as long as 7½ min. At this time the shadow of the moon is always less than 270 km wide traveling at a speed of at least 34 km/min across the Earth's surface.

Astronomers therefore have to plan and organize their travel plans to be in the best location to view a total eclipse, and hope the weather will be kind to them. In the past many total eclipse teams were met with almost insurmountable difficulties; war and shipwrecks were not uncommon obstacles to be overcome. Even when the observers were lucky with the weather, each eclipse seemed to raise more problems than it answered.

Captain Stanyan, observing the total eclipse of May 1706 from Bern, Switzerland is credited with the first recorded observation of the solar "atmosphere". He reported, after watching the Sun emerge from the eclipse "was proceeded by a Blood red streak of Light, from its Left Limb; which continued not longer than 6 or 7 Seconds of Time; then part of the Sun's Disk appeared . . . ". The Astronomer Royal, John Flamsteed (1646–1719), inferred that rather than being an observation of the solar atmosphere, it proved the existence of a lunar atmosphere. Edmund Halley (1656–1742) (of Comet Halley fame), observing the 1715 total eclipse from London, documented not only the streamers radiating from behind the moon—its corona—but also a "long and very narrow Streak of dusky but strong Red Light seemed to color the dark edge of the Moon . . . ". These are what we now know as solar prominences. Again the streamers and "red flames" were attributed to either the lunar or the Earth's atmosphere rather than to the Sun. It's worth noting that if every total eclipse could have been observed between 1610 and 1800 the total duration would have been approximately 400 min, just over 6½ h over a period close to 200 years. Not a lot of time to record the details of the nebulous observed atmosphere!

Alexander Wilson (1714–1786), observing the Sun from Glasgow, Scotland, got further than most eclipse observers of the Sun when he noted that the shape of a circular sunspot penumbra appeared distorted when viewed close to the limb. From this he inferred that the sunspot formed a depression in the surrounding surface, subsequently called the Wilson effect.

Barring these observations, there were no further significant solar discoveries made until the nineteenth century. It wasn't until the 1860s that the actual cause of the red flames that had been observed at various times was finally established by Norman Lockyer (1836–1920) using an early spectroscope. Advances in the scientific understanding of sunspots is addressed in the next section.

1.1.3 The Sunspot Cycle

By noting the visibility of sunspot groups, the longer term 11-year solar sunspot cycle was determined by Heinrich Schwabe (1789–1875). Working from the attic of his home in Dessau, Schwabe had started to observe and record sunspots in 1825. He found that in some years, such as 1828, there were so many sunspots it became difficult to keep track of them. The sunspot numbers began to decline and then rose again to a maximum in 1837. By 1843, after some 17 years of observations, he had gathered enough data to convince himself that the numbers of visible sunspots was showing a pattern of a decade or so between the maxima. He predicted that the next sunspot maximum would occur in 1849 and drop to a minimum some 5 years later. His ongoing results were published annually and confirmed the prediction.

September 1st 1859 proved to be an exciting day for the Victorian solar astronomer Richard Carrington (1826–1875) for reasons related to solar activity. Carrington had been employed as an observer at the Durham Observatory and observed the Sun for many years before joining the Harvard College Observatory eclipse party in 1851. That year's eclipse was visible in Sweden and Carrington made observations of both the corona and the elusive "red flames" during totality. He was enthralled by the event. Was there a connection between the sunspots he had been observing and these red flames? He felt there was, but confirmation had to wait a bit longer. By 1852 Carrington had left Durham and established an observatory in Redhill, Surrey where he set up a 4½″ refractor on an equatorial mount with the initial intent of compiling an accurate catalogue of the positions of the northern stars.

His life was about to change; he heard about Schwabe's work and read that Rudolph Wolf (1816–1893) from the University of Bern had investigated previous records and had managed to extend the sunspot data back to early 1755. It now appeared that the average sunspot cycle was closer to 11.11 years. Carrington was also somewhat surprised to find that there was very little data available prior to that date. This could have been due to the lack of observers—or the lack of sunspots to record. He then visited the library of the Royal Astronomical Society in London to review the historical solar records, where to his amazement he found that the records showed a very casual approach to the subject. There was no fixed frame of reference for the rotational period of the Sun and little agreement on the solar axis of rotation. Carrington decided it would be his role in life to correct these deficiencies and set about organizing a disciplined regular observational project to gather and prepare accurate data on sunspots.

By projecting an enlarged image of the Sun onto a white board and using fine cross-wires at the eyepiece he was able to measure the position of the various sunspots on the solar disk, and also by determining the time it took for the sunspot to drift across the crosswire its approximate size. This became his daily task for the next 11 years.

The historical data showed that the Sun's rotation could be anywhere from 25 to 28 days, and Carrington was determined to find an accurate answer. He assumed that the sunspots were carried along lines of latitude on the solar surface but did not know if the sunspots themselves moved on the surface, like leaves on the surface of a pond. By 1858 Carrington finally realized what was happening: the Sun didn't rotate as a solid body it showed differential rotation, rotating faster at the equator than at the poles. At the equator the rotation had a period of 25 days whereas at higher latitudes it was 28 days, with an average of 27 days. He also noted that at the beginning of a solar cycle the sunspots appeared at higher latitudes then as more spots became visible they were found closer to the equator. This was later confirmed by Gustav Sporer (1822–1895) and become known as Sporer's law. We now recognize it as the Butterfly diagram of the solar cycle (see Chap. 3 for more).

On the morning of the 1st September 1859, Carrington was observing the solar disk as usual and was busy admiring one of the largest sunspot group he had seen, this group extended almost one tenth the solar diameter. Without warning two extremely bright points of light suddenly appeared *"as bright as lightning"*, these brightened even further and became kidney shaped before fading and drifting across the sunspot to disappear some 5 min later. Never had such an occurrence been recorded before. As he continued to observe the sunspot group returned to a more normal appearance. He had just observed a spectacular solar flare which resulted in the most dynamic magnetic storm and Aurora activity ever seen over the next few nights.

The scientific community now had some proof that solar activity could and did impact on the Earth's magnetic field.

Some years before, in 1842, General Edward Sabine (1788–1883) had established one of the first geomagnetic observatories in the world at the Kew Observatory in London. The intent was to measure and record magnetic disturbances, thereby confirming the relationship between magnetic activity and the often seen Aurora. Also at the Kew Observatory, following John Herschel's (1792–1871) pioneering work in photography, Sabine managed to get Warren de la Rue (1815–1889) to invest both his time and money in building a photoheliograph—a specialized telescope with the sole purpose of imaging the Solar surface daily. This refractor type instrument had an aperture of 3.4″ and 50″ focal length and was built by Ross of London. The design included a built-in enlarging lens to produce a final solar image of 12″ on a colloidal wet photographic plate. It was hoped that the images obtained would supplement the data being collected on the magnetic variations and confirm these were in some way connected to the sunspots. Although Kew observatory had not obtained any photographic or visual record of the flare, the magnetic equipment quickly confirmed the sheer scale of the disruption. A magnetic storm of the first degree had just occurred.

The magnitude of the solar flare was enormous and of such significance that it is still referred to as the Carrington super flare or Carrington event.

Carrington also developed the "Carrington rotation"; a system for comparing locations on the Sun over a period of time, thereby allowing sunspots to be identified upon their return after a solar revolution. It is based on a rotational period of 27.2753 days and the Carrington Rotation Number starts from November 9, 1853.

In 1860 De la Rue temporarily relocated the photoheliograph to Rivabellosa, Spain to attempt the imaging of the solar eclipse. He was successful in recording the first images of the solar chromosphere and finally demonstrated that the chromosphere was part of the solar atmosphere and not associated with the moon or the Earth's atmosphere. Similar confirmation images were obtained by Father Angelo Secchi (1818–1878) also observing from Spain. The Kew photoheliograph continued to take daily solar images up to the late 1890s.

As the nineteenth century progressed, other active solar observers like Herschel, Nasmyth, Huggins, Lockyer, Dawes, Wilson, Langley and Young all contributed to the growing knowledge and understanding of the nature of the Sun. It was the birth of solar studies in earnest. This is the period when the texture of the Sun's surface under low magnification was likened to rough drawing paper or curdled milk by the American Charles Young (1834–1908); nowadays we would call this the flocculi network. At higher magnification and steady seeing conditions, this network starts to resolve itself into the granulation, a term first used by William Dawes (1762–1836). These "grains", first noted by James Nasmyth (1808–1890) in 1861 who described them as "willow-leaves", were said to form a "sort of blanket-work formed by the interweaving of such filaments". Secchi thought them to be much smaller and compared them to rice-grains. Young also commented that some surface structures which appeared as "bits of straw" lying roughly parallel to each other, were sometimes noted in and around the penumbral regions of sunspots. The faculae recorded by Galileo were recorded as "looking much like the flecks of foam which mark the surface of a stream below a waterfall". From being an afterthought of astronomers, the Sun now had their full attention, and was inspiring flights of fancy as well as increasing the astronomical community's knowledge.

1.1.4 Photography

While the astronomers at Kew continued their solar photography, the lack of reliable and easy to use photographic material hampered the amateur. The original Daguerreotype (1839) and later wet colloidal glass plates (1850), although very popular, demanded care and attention not to mention very long exposure times. The competitive photographic method developed in 1841 by Fox Talbot (1800–1877) where a transparent paper negative was exposed, developed in silver nitrate, and then contact printed on salt-treated paper to produce a final print was still poorly accepted, even although Talbot's process allowed a single negative to produce multiple prints.

All this was to change in 1880 when gelatine replaced the old colloidal mixture and the Dry Plate Process was introduced. This new process allowed greater flexibility for the user and the short exposure times, seconds instead of minutes, became the catalyst for further camera development. Fast leaf shutters, allowing exposures of 1/5000 s and much improved multiple element lenses were quickly brought to the market. George Eastman (1854–1932) took the popular camera to the next stage with the release of his first mass market Kodak "you press the button, we do the rest" camera in 1890. The Kodak camera made use of a length of rolled flexible film based on a celluloid base coated with silver bromide. This film produced the negative from which multiple image prints could be produced. The success of this film based technology was to go on with improvements to sensitivity and the advent of color film to support the whole of the photographic industry for the next hundred years, only to be superseded in the twenty-first century by the introduction of the digital CCD technology. It is what allowed amateurs to step up their pursuit of solar viewing even further into the realm of imaging.

The invention and adoption of the spectroscope by chemists and astronomers in the mid 1800s added another dimension to the solar investigations, and the work of these early pioneers is discussed later in Chap. 4. The next chapter, however, focuses on the here-and-now of observing the Sun—and how to do it safely.

Further Reading

Clerke, A.M.: A Popular History of Astronomy. Adam & Charles Black (1887)
Young, C.A.: The Sun, Kegan Paul, Trench & Co. (1882)
Clark, S.: The Sun Kings, Princetown University Press, (2007)
Baatz, W.: Photography: A concise history, Laurence King (1999)

Chapter 2

Safe Filters for Solar Imaging

Now that a brief history of solar observing has been given, we are almost ready to move on to the meat of solar observation. First, however, tools must be discussed. This is of great importance for an amateur's outcomes, and it is a safety issue as well. Imaging the Sun with safety in various wavelengths and bandwidths has evolved substantially since the first attempts in the 1800s, and now requires the application of some very precise filter coating technologies. These sophisticated filter construction requirements can lead to expensive solar telescope designs.

Remember at all times, observing/imaging the Sun with any optical aid can be dangerous and can cause severe eye damage. With the proper application of safe filters the imaging of the solar features can be a very safe, rewarding and satisfying pastime. The solar surface presents one of the few dynamic targets in amateur astronomy. The recorded detail can change within minutes, making the task even more challenging, however the excitement of imaging detail like prominences lifting off from the solar surface or the explosions of Ellerman bombs make it all worthwhile.

Over the years solar observers have tried many different solutions to the problem of reducing the solar energy to safe levels for viewing. Some of these solutions were found to be better than others. For instance, notice in this section that welding glass and exposed photographic film don't get mentioned. These materials are not suitable for use by the solar imager and should not be considered at any time, though they have in the past been adopted as methods for visual observers of solar eclipses. This path is not recommended!

The following section reviews some of the significant developments in filter designs for white light and narrowband applications. The current commercial versions of some of these filter designs are detailed in Chap. 3 where the use of the various filters for imaging of the Sun is discussed in detail.

© Springer International Publishing Switzerland 2016
K.M. Harrison, *Imaging Sunlight Using a Digital Spectroheliograph*,
The Patrick Moore Practical Astronomy Series, DOI 10.1007/978-3-319-24874-5_2

2.1   Glass Filters

Colored glass filters have been used in photography and astronomy for the past hundred years. In that time they have been successfully deployed by generations of amateurs to image all sorts of astronomical objects. The early solar observers used smoked glass filters made by holding a plain glass plate over a smoky candle in an attempt to reduce the intensity of the sunlight. Later the use of dark green supposedly solar eyepiece filters became popular. These were intended to be used with small refractors or in conjunction with a Herschel Solar wedge. Unfortunately these are extremely dangerous. There are many reported incidents where the "solar" eyepiece filter has cracked due to the intense heat close to the focus of the telescope. If you do come across one of these small "solar" filters—sometimes unfortunately still found in the beginner's "First telescope" package as a "Sun filter," the recommended action is to take it in your right hand and throw it from a high cliff into the ocean where it can do no more damage — or smash it with a hammer. You will NEVER forgive yourself if you allow it to be used by a novice who subsequently suffers from irreparable eye damage. Nowadays all colored filters are used in conjunction with a full aperture solar film ND filter, a Herschel wedge or an ERF built into the optical train.

The Kodak Wratten number series for defining colored filters was developed by Fredrick Wratten (1840–1926) whose business was absorbed into Eastman's Kodak empire in 1912 and is still commonly used to specify the various filter transmission bandwidths. The Wratten numbering series is not self explanatory and the individual transmission curves need to be reviewed for each application.

Popular Wratten filters used by solar observers/imagers include:

Wratten 80A—Blue filter passing below 5200 Å
Wratten 40/58/60—Green filters similar to Continuum filters, peak transmission around 5300 Å
Wratten 24/25/29—Red filters blocking below 5750Å

Wratten filters are available in both gelatin and dyed in mass versions. Tiffen, Hoya and other manufacturers sell a wide range of quality glass photographic filters in various sizes to suit commercial cameras. Unfortunately the photographic filter sizes used in astronomical equipment are non standard and it can be difficult at times to find M48 threaded filters suitable for mounting into 2″ astronomical fittings. Also, the surface accuracy of the glass substrate is not necessarily of the highest quality. Most of these photographic filters are much less than the ¼ wave surface accuracy required in precision optics.

Schott manufacture a small but important range of dyed in the mass glass filters. Traditionally the red RG610 and RG620 glass have been used as energy rejection filters (ERF) in solar telescopes. Schott also manufacture some interesting special glasses, KG5 is a Schott Infra-red (IR) absorbing glass which can be used successfully in solar imaging to suppress any IR leakage from the narrowband filters. The standard sizes seem to be 25 mm diameter or 50×50mm

2.2 Neutral Density (ND) and Optical Density (OD)

Originally for use with cameras and photographic film, the Neutral Density (ND) filters allowed the amount of light received by the film to be reduced without changing the aperture stop. The reduction in light transmission then allowed the photographer to have more control over the exposure time and depth of field. All ND filters absorb equally across the whole visible spectrum. The ND number of a Neutral Density filter defines the attenuation (reduction in transmission) and Optical Density (OD) of the filter. The ND and OD ratings are identical. ND numbers are additive i.e. a combination of a ND2 and an ND3 filter will be equal to a ND5 filter.

The optical density (OD) is related to the transmission by the following equation:

$$T = 10^{-OD} \times 100 = \text{percentage transmission.}$$

$$OD = -\log(T / 100)$$

Example: If an ND 0.9 and a ND 2 filter are combined, what is the final percentage transmission?

$$ND = 0.9 + 2.0 = 2.9$$

$$T = 10^{-2.9} \times 100 = 0.00126 \times 100 = 0.126\%$$

ND/OD rating	Transmission (%)
ND 0.9 (OD 0.9)	12.5
ND 2.0 (OD 2.0)	1
ND 3.0 (OD 3.0)	0.098
ND 3.8 (OD 3.8)	0.016
ND 4.0 (OD 4.0)	0.01
ND 5.0 (OD 5.0)	0.001

In solar imaging the ND3.8 filter is regularly used for white light; ND5 is the recommended rating for extended visual observing.

All colored glass filters work by absorbing the light and energy from the wavelengths that are not transmitted. This energy absorption results in the filter heating up which can cause optical distortion or in extreme cases breakage. For solar work the modern multi-coated reflective filters are safer.

2.3 Multi-coated Filters

Modern developments in multi coating of glass filters has resulted in some very effective anti-reflection coatings (AR) being designed and made available. These AR coatings are now widely used on quality optical and astronomical lenses and associated equipment. The same techniques have also allowed narrow band dichroic filters, which block or transmit colors based on wavelength, to be constructed.

The popular Induced Transmission Filters (ITF) narrow band filters we see today are usually constructed by vacuum depositing a very thin, $\lambda/4$ thick metallic film on a glass substrate. Further dielectric films of $\lambda/4$ thickness and having specific design refractive indices are deposited to present an interference layer which suppresses and reflects all the wavelengths other than the design bandwidth. In these interference filters, light traveling from a lower index material will reflect off a higher index material; only light of a certain angle and wavelength will constructively interfere with the incoming beam and pass through the material, while all other light will destructively interfere and reflect off the material.

Soft coated multi-coated ITF filters are constructed by applying the sensitive coatings to two substrate plates. The mating edge circumference of each plate is cleared of coating and the sandwich is laminated together with moisture resistant epoxy glue. The glue gives a glass-epoxy-glass bond at the edge circumference to seal the filter and prevent moisture ingress. After final cutting to the finished size additional sealant may be added to the edge to further improve the moisture barrier and give added protection to prevent premature failure of the coatings. A similar process can be applied where the epoxy is only applied around the external edge. In this case there is no epoxy present in the optical path. This is the preferred construction of ITF filters used for solar work.

Hard sputtered filters have the multi-coated layers applied directly to one surface of the substrate and the choice of the film material and the application process provide a hard strong and durable external coating finish which can be cleaned without damage.

Typically the final coating thickness will be 10–20 µm and can take many hours to manufacture. The complexity, quality control, and manufacturing time adds to the cost of this filter design. The successful Baader D-ERF filter and the BelOptik Tri-ERF (CaK, Continuum, and Hα) are examples of this type of filter construction.

One limitation of all these dichroic interference filters is the limited field angle acceptance. As the angle of incidence increases through the filter the central wavelength of the bandpass will shift towards shorter wavelengths (i.e. towards bluer wavelengths).

2.4   White Light Solar Film Filters

Back in the 1970s Roger Tuthill (1919–2000) was the first to supply an aluminum coated Mylar filter, his "Solar Skreen", to the astronomical market. This was well accepted by the amateur community as a safe solar white light filter. In a 1981 review by Dr. Chou of the School of Optometry, University of Waterloo, Toronto, it was favorably compared to the then-available aluminum coated premium Questar glass filter. Although providing a very safe ND level the final image quality was marred by the optical performance of the Mylar which presented a high degree of scatter due to internal strains and the grain structure of the material. The introduction of the Baader "AstroSolar" safety film in 1999 was a major step forward. This is a proprietary polymer film which has a surface accuracy of better than 1/8 wave and provides ND5 (neutral density factor 5) attenuation. The surfaces of the film are then ion implanted and coated with aluminum on both sides. Subsequent testing showed this solar film provided twice the optical quality of the aluminized Mylar and four times that of some inexpensive glass solar filters. It is available for DIY installation in ND3.8 (imaging) and ND5 (visual) variants and sheet sizes up to A3.

Only 1/10,000 of the incident light is allowed to pass through the filter. This complies with the international recommended solar protection level of ND3 or greater. The surface quality of the Baader film ensures that the solar detail visible and images recorded are not compromised. The final results will be limited by the seeing conditions and the optical quality of the instrument, but this remains a highly recommended solution.

There are other options, as well. The polymer/glass filters available from Thousand Oaks, JMB, Orion, Seymour and others are all safe to use. Check out the optical quality before making a final decision.

2.5 Herschel Solar Wedge

Invented by John Herschel in the 1800s, this solar filter externally looks like a star diagonal with an open section at the rear. Internally a plate of optical glass shaped in thickness to a wedge (a 10–30° taper) replaces the conventional reflective mirror and allows only 4 % of the incoming light to be reflected from the front surface and directed to the camera/eyepiece; the wedge shaped body then deflects the rest of the light and heat out of the rear of the diagonal housing. To reduce the final incident energy to safe levels the Herschel wedge MUST be used with a secondary neutral density (ND3) filter, positioned between the wedge and the camera/eyepiece. Likewise, any supplementary filters (i.e. Continuum etc.) must also be placed after the wedge to reduce the likelihood of thermal failure. Although this Herschel wedge is safe to use on all refractors, some manufactures do however, suggest limiting the telescope aperture to 125 mm or less to minimize any overheating. Most amateurs who have used the Herschel wedge believe the results are superior to any other design of white light solar filter. Baader, Lunt, Intes and others supply Herschel wedges in 1.25 ″ and 2 ″ sizes to fit most refractors.

2.6 Chromosphere Filters

The solar filters discussed so far only allow imaging of the photosphere. The structure of the Sun's atmosphere will be discussed in the next chapter, but those details are of no use without a proper filter. To image the upper regions of the solar atmosphere (or chromosphere) where we can see filaments and prominences we need to use very specialized, extremely narrowband filters usually less than 1 Å wide. These filters are typically designed for specific wavelengths associated with the solar emissions of Calcium and Hydrogen.

Using the narrowband filters or a digital spectroheliograph (SHG) will open up a whole new dimension in solar imaging. The dynamic structures in and around the sunspots and active areas can become targets with this filter, from features like extended filaments and prominences of material to light bright plage areas around sunspots and the overlying network of focculi cells throughout the height of the chromosphere.

Since the 1920s astronomers have investigated various optical designs which could be used for the construction of safe and efficient solar narrowband filters. The Lyot polarized filter and the re-interpretation (for solar observing) of the earlier Fabry–Perot filter design were quickly adopted by the professional astronomers. More recently manufactures like Daystar, Coronado and Lunt have refined the Fabry–Perot concept and made quality Hα filters available to the amateur. The Lyot and Fabry–Perot designs are discussed in the following sections. The actual commercial solar filters being used by solar observers are reviewed and discussed in Sect. 3.7.1.

2.6.1 Lyot Filter

Bernard Lyot (1897–1952) a French engineer, worked at Meudon Observatory after the First World War and became expert in the field of polarization and monochromatic light. In 1933 he published his paper on the design of his narrow-band Lyot filter. This filter was constructed from a series of quartz (or Iceland Spar) plates which were separated by sheets of polarizing film.

The general arrangement of the Lyot filter is shown in Fig. 2.1. The polarizing films (with the planes of polarization parallel to each other) are shown as P_1, P_2 through to P_7) and the varying length quartz plates are numbered 1–6. Each plate in the filter series is double the length of its predecessor. The lower part of the diagram shows the various interference fringes produced by the succession of plates and shows only the two monochromatic wavelengths 6370 and 6563 Å will pass through the filter assembly.

The length of each plate required is calculated to produce a series of interference fringes which, upon final assembly, results in the isolation of the target wavelength required. The unwanted secondary wavelength (in the example illustrated 6370 Å) can be blocked by adding a suitable conventional blocking filter. This Lyot filter is capable of providing relatively wide field images of the solar disk in the design wavelength i.e. Hα

To ensure the dimensional stability of the plates, the Lyot filter is designed for use at a specific temperature and assembled in a temperature controlled enclosure. Pettit (Amateur Telescope Making 3, p. 416) gives full construction details of a

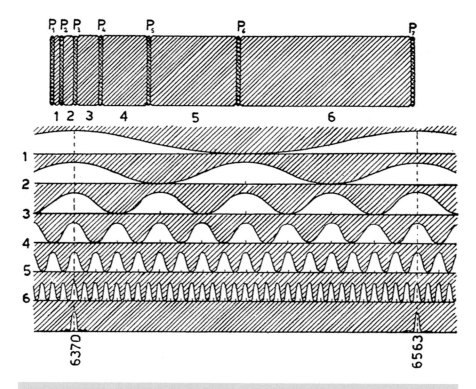

Fig. 2.1 Diagram of the monochromatic polarizing Lyot filter (Lyot)

quartz Lyot Hα filter and notes that a 1 °C change in operating temperature will shift the centerline wavelength by 0.71 Å.

The Lyot filter was widely used in the early part of the twentieth century by most of the professional observatories around the world, utilizing their available telescopes. This allowed them to conduct extensive solar observing (and imaging) without the added costs (and complexity) of building and maintaining a large "tower" type solar installation.

Due to the build accuracy and quality of the optical arrangement required to perfect the Lyot filter, it never took off with the amateur community who continued to work with other options like the SHG/scope.

2.6.2   Fabry–Perot (F–P) Filter

In 1897 the French scientists Maurice Fabry (1867–1945) and Alfred Perot (1863–1925) published details of their interference filter, the F-P interferometer or etalon. The construction of the filter was based on two highly reflective glass plates held exactly parallel to each other. The fringes produced in the transmitted light after multiple reflections between the surfaces the spacing between the plates and the field angle θ (see Sect. 6.4.4) of the incident beam give rise to a series of interference rings. Each ring is a separate spectral order and it is usual to operate the etalon where the central order $(\theta = 0)$ is isolated. The spacing between these fringes is the free spectral range (FSR) $(\Delta\lambda)$ and is related to the spectral order (m), the spacing between the plates (d) the refractive index of the spacer material (n)—usually air this gives:

$$\Delta\lambda = \lambda / m = \lambda^2 / 2nd$$

As the number of spectral orders (m) is usually large a secondary blocking filter is used in conjunction with the etalon to isolate the target CWL and suppress all the others. The finesse of the etalon is a number (N) given by:

$$N = \delta\lambda / \Delta\lambda = \lambda^2 / \delta\lambda\, 2nd$$

The finesse is only related to the quality and reflectivity of the etalon plates and typically is 30–50.

Finally the spectral resolution (R) of an etalon is given by:

$$R = \lambda / \delta\lambda = N\lambda / \Delta\lambda = 2Nnd / \lambda$$

The etalon filter will always be used in conjunction with an energy reduction filter (ERF). The ERF can either be a full aperture filter or a sub-diameter filter placed in the optical path. The incident beam to the etalon should be parallel and close to collimated. When sub-diameter etalons are used in solar telescopes a Barlow (or telenegative) lens is positioned in front of the etalon to give the required parallel beam. A re-imaging lens is placed behind the etalon to refocus the image.

The resulting multiple orders from the etalon appear, when viewed with a spectrograph, as a series of "comb" like peaks across the spectrum each separated from the other by the free spectral range (FSR). When considering a bandwidth of <1 in the Hα region the separation between successive peaks is usually around 10–12 Å. A narrowband sorting (or blocking) filter is placed after the etalon which only allows the central wavelength (CWL) to be transmitted. All the other peaks are suppressed and blocked as shown in Fig. 2.2. The blocking filter may also be protected by an IR induced transmission filter (ITF) to reduce thermal effects.

The size of the blocking filter required to give a non vignetted field large enough to accommodate a full solar disk depends on the effective focal length of the telescope. The minimum size is approximately 1/100 the focal length i.e. a 500 mm fl system would need a 5 mm diameter blocking filter.

The Jacquinot spot (named after Pierre Jacquinot (1910–2002)) or more commonly called the sweet spot is the region of the image where the most uniform monochromatic field is recorded. The Jacquinot, spot is defined to be the region over which the change in wavelength does not exceed $\sqrt{2}$ times the etalon bandpass.

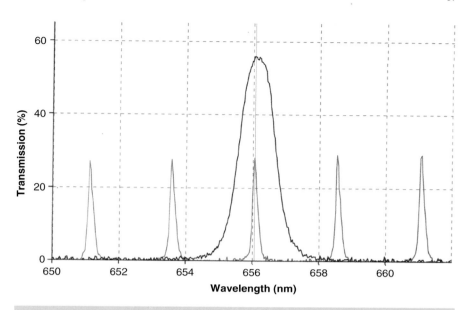

Fig. 2.2 F–P etalon transmission and blocking filter curve

The etalon type solar filter came into common use by the amateur in the late 1990s when DayStar and then Coronado released their solar Hα range of filters.

There are two popular types of etalon construction that the amateur will come across—air spaced and mica spaced.

In the air spaced design the etalon plates are separated by a series of small thin spacers optically bonded to the outer edge of the plates.

Tuning and adjustment of the CWL requires the etalon assembly to be either tilted within the support housing, to slightly increase the angle of the incident beam and hence the effective spacing between the etalon plates or enclosed in a chamber where the internal air pressure can be altered by means of a piston arrangement. The pressure increase in the chamber changes the refractive index of the enclosed air and hence the effective spacing between the plates. This is what is meant by "tilt tuning" and "pressure tuning".

For solar Hα filters, manufacturers design the etalon spacing to give a central wavelength (CWL) just to the red side (red wing) of the Hα wavelength at approximately +1Å . As the filter is tuned, the central wavelength moves towards the blue wavelengths, first reaching the Hα core then further into the blue wing and finally to the surrounding solar continuum (see Sect. 3.7.).

Air spaced tilt tuned etalons are commonly used as both front aperture solar filters and internal assemblies. The external etalons in larger sizes (40 mm and above) can be used on any refractor whereas the internally mounted (with apertures up to approximately 30 mm) are normally supplied pre-fitted to a solar telescope system where the effective focal ratios are f8 or above. The air spaced etalon is also a favorite for double stacking (DS) where two etalons are combined in the optical system (either as double external etalon or one external followed by an internal etalon). The double stack configuration effectively reduces the bandwidth to the product of the individual etalon bandwidths i.e. a DS of 0.7 Å etalons would give a $(0.7 \times 0.7) = 0.5$ Å outcome. DS also has the advantage of improving the finesse of the bandwidth, which in turn can reduce the parasitic (photospheric) light in Hα imagining.

The construction of the mica spaced etalons is more complex. Large crystals of mica are cleaved (split along the natural crystal plane) to produce a very flat constant thickness thin sheet. The thickness of the mica will determine the free spectral range (FSR) and also the temperature at which the etalon will be on band. Typical thicknesses range from 10 to 30 μm. Both sides of the mica plate is then given a highly reflective coating and bonded to glass substrates. As the mica is a birefringent material producing both normal and extra-normal polarized light, the mica is then sandwiched between two circular polarizing filters which suppress the extraordinary light and reduces unwanted reflections from the filter assembly.

An infra-red (IR) suppressing ITF filter element and a blocking filter are incorporated to into the filter assembly which is finally mounted inside the housing and sealed with external glass windows. To tune the etalon to the required central wavelength (CWL) the filter is surrounded by a temperature controlled oven which can heat or cool the filter. The temperature of the filter is usually maintained within 0.1° of the target temperature. A change of temperature of 9.5 °C will shift the CWL by 1 Å.

All the commercial narrowband Hα filters currently available to the amateur are based on the etalon filter/blocking filter design concept. Any of the filters above are suitable for getting started or upgrading, and one at least is necessary if one desires to move into this area of observing. As for what you'll be observing—the following chapter goes in-depth on the chromosphere.

Further Reading

Robertson, J.K.: Introduction to physical Optics, Van Nostrand, 1950
Jenkins, F.A and White, H.E.: Fundamentals of optics, McGraw-Hill, 1965
Hadley, L.N., Dennison, D.M.: "Reflection and Transmission Interference Filters," J. Opt. Soc. Am. **37**, 451-453 (1947)

Webpages

http://www.astrosurf.com/viladrich/astro/instrument/solar/FP.htm
http://www.astro.umd.edu/~veilleux/mmtf/basic.html
http://www.company7.com/astrophy/options/apasolar.html

Chapter 3

The Sun and Its Atmosphere

As a reminder, observing the Sun can be dangerous. Even extended naked eye observations can irritate the eye and possibly cause temporary sight impairment. The internationally recommended minimum safe protection level across the visible spectrum for extended exposure is ND3 (Neutral density #3). All the commercial white light solar filters available to the amateur comply with this guideline.

Nevertheless there is always the risk of accidental exposure. All filters must be securely fixed in place at all times. Never leave a solar telescope unattended; as we know, young children can be very inquisitive and don't always ask first before acting.

Take care at all times and an exciting world of discovery awaits.

3.1 Introduction

This chapter provides the basic data on the Sun and how the amateur can safely image/observe the solar surface in different wavelengths of light. For both white light or photosphere and narrow band or chromosphere observing the significant features which can be recorded will be discussed. The commercial versions of the solar filter designs previously discussed in Chap. 2 are highlighted and illustrated with examples that clearly show the results which can be obtained using them.

The significant effects of the solar magnetic field, which can be studied with the digital SHG, are also explained.

© Springer International Publishing Switzerland 2016
K.M. Harrison, *Imaging Sunlight Using a Digital Spectroheliograph*,
The Patrick Moore Practical Astronomy Series, DOI 10.1007/978-3-319-24874-5_3

3.2   Our Nearest Star

The Sun is our nearest star and the only star which can be observed as a disk, allowing us to observe surface detail. It's a middle aged G2V star (see Chap. 5 for more on what this means) of average size, with a mass of 1.989×10^{30} kg and a diameter of 13.92×10^5 km. Although only an average sized star, the Sun still contains 99.9 % of the total mass of the solar system. Even with the extreme mass of the Sun, the average density is only about ¼ that of the earth and a specific gravity of 1.41. The escape velocity is 618 km/s on the Sun compared to only 11 km/s on the earth. Any material ejected by flares or eruptive prominences exceeding this velocity will flow away from the Sun into the solar wind, but material at a slower pace gets inexorably pulled back by the star's gravitational force.

The Sun's core (which is approximately 25 % the solar diameter) has a temperature of around 15 million degrees. This amazing temperature is maintained by nuclear fusion, where hydrogen nuclei (protons) are converted into helium nuclei. During this reaction a small but significant mass loss occurs which is released as energy. Every second the Sun converts approximately 700 million tons of hydrogen into helium, with only 0.7 % being released as energy. This may sound like a lot of weight loss, but it only represents 1/4000 of its mass being lost during the current lifetime of the Sun. These core reactions give rise to intense radiation, initially as high energy gamma rays then as absorption and re-emission in the surrounding material occurs, as X-rays.

As the energy continues to diffuse slowly outwards from the core through the radiative zone the photon energy decreases, giving us ultraviolet (UV) light and then, once it is finally at the surface, visible light.

It is only when the temperature falls below 2 million degrees at about 70 % of the solar radius that the opacity of the material increases and the energy starts to be transported via convection. This is the outer region of the Sun's core, where material rises through the convection zone layer to be finally seen as the granulation in the photosphere. This is also the region which generates the solar magnetic field.

The photosphere defines the Sun's visible surface. This region at the top of the convection zone presents a mostly "solid" surface and is only 200 km deep with an average surface temperature of 5780°K. The lower parts of the photosphere are actually at a slightly higher temperature, 6370°K which decreases with height to 4950° at the top of the photosphere. When the Sun is imaged in so-called white light across the full visible wavelengths from 4000 to 7000 Å, it presents a crisp edge with no indication of haziness. The only suggestion of depth is indicated by the limb edge darkening effect where we view through a deeper, and cooler, region of the photosphere. It is in the photosphere that the effects of the solar magnetic field start to become obvious. The Sun appears to be a highly magnetic star, the effect of the magnetic fields dominate the visible features and give rise to an 11 year cycle of solar sunspot activity. We will discuss later the photospheric detail which can be imaged by the amateur.

Immediately above the photosphere is the region known as the chromosphere. Transparent and invisible in white light images, the chromosphere extends from the upper heights of the photosphere at around 200 km for at least 10,000 km. This extended part of the solar atmosphere was first observed during the occasional total solar eclipse. The bright transient flash spectrum seen at these total eclipses, when the moon just leaves a very narrow arc of the Sun exposed, gave rise to the name reversing layer. The very limited time between the flash spectrum appearing and the brighter photosphere returning the spectrum to the normal absorption spectrum indicated that this layer could only be a few hundred kilometers deep. The emission lines observed appeared to match the wavelengths of the prominent solar absorption lines already noted (i.e. hydrogen), but reversed into strong emission lines. This can only occur if the gaseous material in this region is at a lower temperature than the underlying photosphere, due to Kirchhoff's laws, which will be discussed at greater length in the next chapter.

Figure 3.1 shows the solar environment and the various regions (and features) of interest to the amateur. The photosphere and chromosphere are the most fruitful areas of challenge for the amateur solar imager and by using the correct equipment we can obtain spectacular images of active areas, sunspots, filaments and prominences.

The temperature of the chromosphere, initially the same as the photosphere's approximate 6000°, falls rapidly to around 4000° at a height of 500 km. This gives rise to the reversing layer before steadily rising again to 6000°, between heights of 1000 and 2000 km. Beyond this region, in the Transition zone, the temperature dramatically rises from 20,000 to beyond 2,000,000° as we reach the solar corona (see Fig. 3.2).

The varying temperature and pressure conditions within the upper photosphere and the chromosphere also control the ionization conditions of the atoms which in turn give rise to the absorption features we see in the solar spectrum, discussed further in Sect. 5.3. Observing the many and varied solar chromosphere features such as filaments, prominences, flares demands the use of narrowband filters. Prior to the twentieth century only a few lucky observers like Hale and Deslandres had actually imaged the extent of solar chromosphere. The early observations of the extended chromosphere during total eclipses only allowed brief views of the limb spiculation layer and larger prominences. Now all the features can be recorded by the amateur using the digital SHG or the specialized narrowband filters mentioned earlier in Chap. 2.

At the extremes of the solar atmosphere and extending well beyond the solar surface is the corona, which can be recorded millions of kilometers into space and is associated with the Solar wind which impacts on the Earth. The impact of the Solar wind on the earth's magnetic field can give rise to critical power surges in electrical distribution systems, as well as causing the beautiful Aurorae at the poles. The corona is only visible during a total eclipse and by using specialized chrono-graph telescopes.

The energy we receive from the Sun can be considered a continuous series of electromagnetic waves (see Fig. 3.3). These extend from the short wavelength but

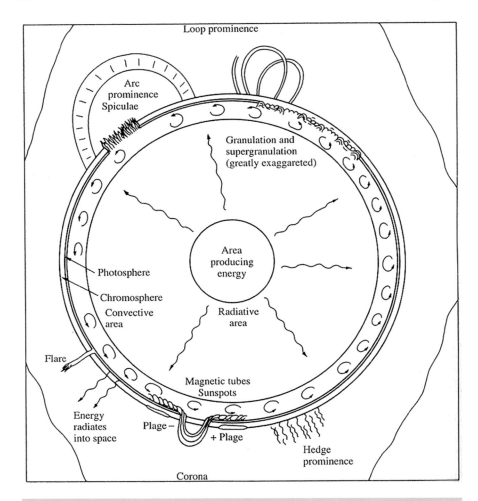

Fig. 3.1 Interior and surface of Sun (after Van Zandt)

high energy Gamma and X-rays through to the visual spectrum—blue to deep red—and beyond into the infrared and radio waves.

The energy distribution is not uniform across all wavelengths and was first noted by Fraunhofer to peak in the 5500 Å wavelengths, in the yellow-green portion of the spectrum. This will be discussed in Chap. 5 when we look in more detail at what the solar spectrum can reveal to us.

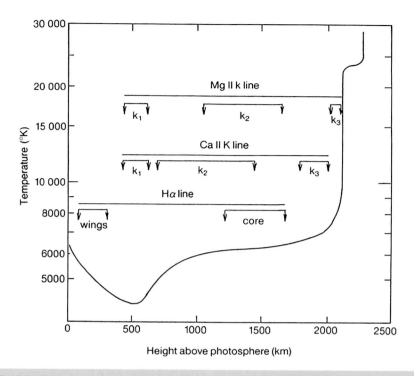

Fig. 3.2 Chromosphere regions recorded at different wavelengths (Phillips)

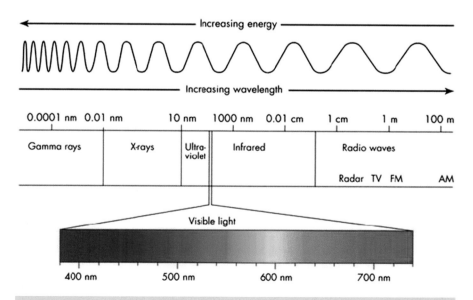

Fig. 3.3 Electromagnetic spectrum (WIKI)

3.2.1   The Magnetic Sun

Since the time of the 1859 Carrington Event previously mentioned and the subsequent magnetic storm recorded at the Kew Observatory and elsewhere, the question of the how the Sun's activity interacted with the Earth's magnetic field was a matter of debate by astronomers, particularly when it came to the possible impact of the Sun on the Aurorae. Elias Loomis (1811–1889) an American mathematician at the University of New York investigated the magnetic storms associated with Carrington's flare, the first being recorded on the 28th August with a later, even stronger storm recorded on the 2nd September 1859, and found the effects not only caused the largest widespread aurora activity ever seen, but made a dramatic impact on the telegraph systems around the world. Balfour Stewart (1828–1887) the Director of the Kew Observatory later published a paper which stated the following:

> if it be true that the spots on the surface of our luminary (or action connected with these spots) are the primary cause of magnetic disturbance, it is to be hoped, since the study of the Sun's disk is at present a favourite subject with observers, that ere long something more definite may be known with regard to the exact relationship that subsides between these two great phenomena.

Loomis and others concluded that somehow there was an electrical connection of some sort, previously unimagined, between the Sun and the Earth's magnetic field. Sabine also felt there was a relationship between the sunspot activity and the magnetic storms, since the aurora only seemed to appear when large sunspots were visible. American George Ellery Hale was the first solar observer to record a solar flare in light of hydrogen using his spectroheliograph. Observing on the 15th July 1892 he imaged a pair of sunspots near the center of the Sun; over a period of 30 min, his spectroheliograms showed the fade from a bright white streak and terminal brilliant spot to a diffuse cloud sitting above the sunspots. An explosion of tremendous energy had just been observed, and similar to the previously recorded flares the effect was felt on the Earth the following day when magnetic storms disrupted the telegraph systems. The evidence was mounting that there was definitely some connection between events on the Sun and the Earth's magnetic field.

E. Walter Maunder (1851–1928), head of the solar department at Greenwich Observatory, had also been closely investigating the sunspot magnetic storm relationship. His review of visible sunspots, their size and location at the time of heightened magnetic disturbances, caused him to conclude that there was a definite pattern of 27 days between events—which matched the rotational period of the Sun. The activity seems to point to a heliocentric position rather than any visible sunspot, large or small. In parallel to these investigations, Maunder's wife Annie by now an accomplished photographer noted that the shape of the extended solar corona recorded during the recent total solar eclipses (1898 and 1901), also appeared to be influenced by the sunspot cycle. At times of the solar maximum the coronal streamers are much more active and appear to extend further beyond the

Sun, whereas at minimum the corona is much more subdued. Maunder asked himself whether it could be that the solar magnetic field was somehow erupting through the solar surface and causing the sunspots. With the coronal activity, perhaps those powerful streamers extended as far as the Earth where they caused the magnetic storms.

When these results were presented to the Royal Astronomical Society (RAS) in 1904 they were initially ridiculed, and it was not until early 1905 at a subsequent RAS meeting that Professor Joseph Larmor (1857–1942) from Cambridge advised the gathering of his researches into electricity and reminded them of the recent discovery of the electron which could transfer electrical charge in a vacuum as seen in the cathode ray tube, put some weight behind Maunder's conclusions.

Johann Hittorf (1824–1914) had in the 1860s shown that two electrodes in a glass vacuum tube caused an electrical charge to build up on the positive electrode. He also determined that the energy beam between the electrodes travelled in straight lines. These were later called cathode rays. This proved to be a fruitful area of research for Joseph John Thomson (1856–1940) working at the Cambridge laboratories. His experiments lead to the discovery, in 1897 that the cathode rays were in fact negatively charged particles and later that the particles were sensitive to electrical and magnetic fields. The direction of the particles could be altered by strong electromagnetic fields. It was quickly established that the electrons as they were now called were much smaller than the atom and must therefore be a sub atomic particle. The atomic structure as we now know it of an atomic nucleus surrounded by electrons was defined.

Astronomers gradually came to terms with the fact that the corona was in fact a manifestation of the solar wind a continuous stream of highly charged particles being emitted by the Sun. The source of these particles is the magnetic fields originating with the Sun's convection zone as the hot plasma (made up from electrons protons and ions) circulates between the radiative zone and the photosphere. We see similar effects then coils of wire wrapped around an iron core which when energized induce a magnetic field into the iron. Similarly when a coil of wire rotates inside a magnetic field an electrical current is induced—the alternator or dynamo in your car is a good example.

The differential rotation of the Sun, discussed in Chap. 1, causes the magnetic fields formed by the plasma material circulating above the solar core to act in a similar manner to a series of giant dynamos. These magnetic fields are distorted and stretched in longitude and could contribute to the "butterfly diagram" distribution of sunspots as well as the bipolar nature of sunspot groups (Sect. 3.5) The net effect is that the magnetic field forms ropes or twisted tubes which inhibit convection and produce zones of lower than normal temperature. The sunspots themselves were associated regions where these complex twisted tubes within the magnetic field break through the photosphere causing localized cooler regions.

The solar flares are a result of the collapse the local magnetic field releasing the bound plasma and giving rise to a dramatic surge of energy which is felt throughout the solar system. The issue of why the Sun's magnetic field appears to reverse every

22 years still remains a mystery, however. This and some of the other secrets the Sun still holds is an ongoing area of research for professional astronomers.

It was Hale who in 1905 established the solar telescopes at Mt. Wilson to take advantage of the better seeing conditions at the altitude of 1700 m. Initially the large horizontal Snow telescope built by Hale with a grant of $10,000 from Miss Helen Snow as a memorial to her father and originally located at Yerkes Observatory in Chicago, equipped with a series of spectroheliographs, was relocated to the mountain, but the results although impressive did not satisfy Hale. He felt the ultimate answer lay in the tower solar telescope design. This design placed the two mirror cœlostat (see Sect. 11.1) well above the ground and the vertical arrangement minimized any temperature effects. In 1907 the 60 ft tower telescope fitted with a 12″ objective, focal length 60′ was commissioned and an even larger 150 ft solar tower telescope using a 12″ objective, focal length 150′ to feed a Littrow spectrograph which could be configured with a 75′ or 30′ collimating lens was constructed in 1909. This arrangement when used with a Michelson grating provided an extraordinary dispersion and resolution. Dispersions as high as 0.1 Å/mm, the equivalent of 0.00056 Å/pixel, are mentioned in Hale's papers.

With these large tower telescopes Hale commenced his serious investigations of the solar magnetic fields. His confirmation of the Zeeman effect (Sect. 5.5) in 1908 allowed accurate measurements to be obtained for the magnetic field strength across the Sun and later with the introduction of magnetography, the bipolar nature of the sunspots. This, like his earlier work on the chromosphere, opened up a new avenue of solar research which is continued today at many observatories around the world.

3.3 Measuring Positions on the Solar Disk: Heliographic Coordinates

The telescopic observations from the past 400 years eventually confirmed the regular sunspot cycles, the axis of rotation which is inclined by 7° 15′ and the average rotational period of 27 days. From the earth's orbit, at an average distance of 149,600,000 km, the solar diameter appears to be 32 arc min. By comparison, the Earth viewed from the same distance would appear as a globe only 17.6 arc seconds in diameter. In terms of scale, this makes the Earth a small disk 1/109 the solar diameter.

As the solar axis of rotation is inclined by 7.25°, the apparent position of solar features like sunspots as viewed from the earth will vary during the solar day, and also with the time of year. Early visual observers plotted the features on a set of pre-prepared Stoneyhurst Disks (Fig. 3.4). These measurement disks were first developed at the Jesuit Stoneyhurst Observatory in England in the late 1800s and thought to be due to the work of Walter Sidgreaves (1837–1919). The series of disks produced showed the equator and the position of the solar north pole

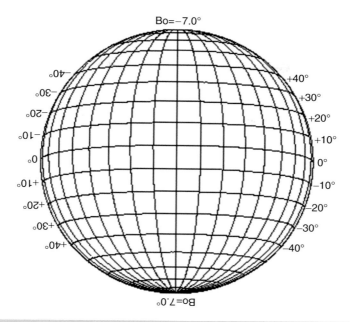

Fig. 3.4 Stonyhurst disk B_o −7 (P Meadows)

(P = angle between the pole and the meridian (±26.4°), B_o = the heliographic latitude of the disk center (±7.25°)) for each month of the year. Daily B_o and P positions are given in many of the available astronomical annual handbooks.

Today there are graphic programs available which can prepare a daily grid of heliographic coordinates that can also be cut and pasted and superimposed on the solar image. The TiltingSun-G software (Fig. 3.5) from Cowley is an excellent example.

These overlays, when used in conjunction with the Raben daily maps described in more detail later, can quickly ensure the orientation of the image is correctly matched.

Figure 3.6a shows the photospheric surface in white light. This image taken on 20 April 2015 displays the typical edge limb darking and the extent of the visible sunspot groups and associated brighter faculae. The identification and annotations of the active areas have been taken from the Raben map (Fig. 3.10) of the same day. Figure 3.6b shows the same image overlayed by a latitude/longitude grid.

On the Raben maps, the preceding edge of the solar disk is shown as a drift alignment mark. This is the point on the disk which will appear first in a camera when the tracking drive is switched off. Moving the telescope in Declination, say towards the North, the last part of the solar disk to be seen will be the northern edge.

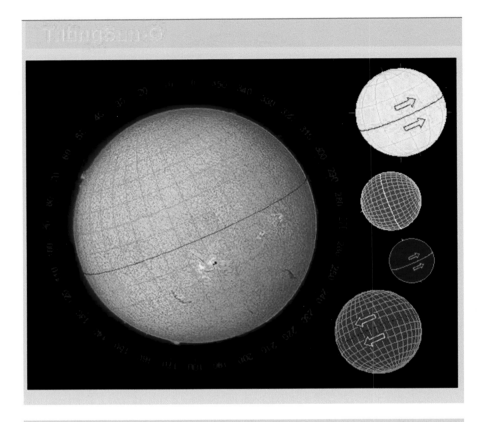

Fig. 3.5 Screen dump from Tilting Sun software (Cowley)

3.4   Imaging the Photosphere

3.4.1   White Light Imaging: Safe Filters

The solar surface or photosphere can be imaged in white light, meaning all visible wavelengths. Amateurs normally make use of small refractors or standard reflectors/SCT fitted with wide bandwidth white light filters to do this. These instruments can be made safe for solar viewing and imaging by fitting a suitable filter which totally and securely covers the front aperture. Remember how important ensuring safety is for this step! A sub-aperture mask can be prepared which blocks the light and only provides a reduced diameter opening for the solar filter. The seeing conditions during the day usually don't favor larger aperture telescopes and the sweet spot aperture for solar viewing/imaging seems to be around 120 mm. This allows an effective sub-aperture mask to be fitted to larger scopes with little impact on the end results.

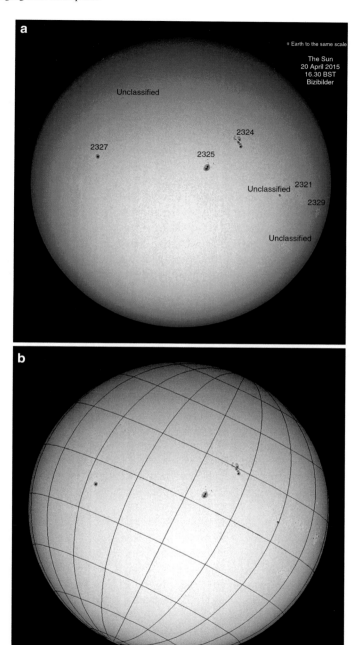

Fig. 3.6 (a) *White* light solar disk (Bisgood). (b) *White* light disk with grid overlay (Bisgood)

One of the most cost effective and most popular designs of solar filter is the AstroSolar film produced by Baader. Although the solar aperture filter will give good safe views of the Sun there are other solutions available to the amateur. A Newtonian reflector with the reflective coating stripped from the primary mirror combined with a neutral density (ND3) filter in the focuser has been used successfully. The uncoated primary reflects only 4% of the incoming light, the balance passes through and out of the telescope, the addition of a ND3 filter ensures that the final light intensity is reduced to a safe level.

For solar observers with refractors the ultimate arrangement is to use a Herschel wedge in combination with an ND3 filter. The white light solar detail of sunspots, faculae and surface granulation can all be recorded with these filters; however, adding supplementary narrow band color filters can further enhance the contrast and definition. The most popular secondary solar filter is the Baader Continuum filter. This is specifically designed to isolate a narrow region of the solar continuum around 5400 Å where the eye is most sensitive. Some amateurs have taken advantage of their available night time filters and added at various times light pollution filters, OIII and DSO Hα filters to their solar filter train to improve the clarity of the photosphere, with mixed results. If you have such filters available, there's no reason not to try them.

3.4.2 Solar Imaging: Camera Selection

For casual white light imaging (using the safe filters detailed above) everything from webcams, DSLR's through to mobile phones have been used. The DSLR, although not commonly recommended for solar imaging, due to the Bayer matrix (see later), has proved quite capable of producing creditable white light full disk solar images. Used with telescopes having effective focal lengths around 1500 mm, or short focus telescopes plus Barlow lens, provides a good image scale and presents a full disk image of approximately 15 mm diameter. Figure 3.6a shows a good DSLR example.

For detailed close-up images around active areas and sunspots a fast frame mono camera gives the opportunity of recording an AVI which can then be processed to minimize the impact of seeing conditions and improve the spatial resolution. The Point Grey or DMK series are examples. Each camera has its own imaging acquisition software. Most new astrophotography programs, like AstroArtV5 etc. are very versatile and can control many different camera platforms from DSLR to CCD cameras. The images once converted to .bmp files can then be contrast enhanced and sharpened to extract the maximum information.

With the fast frame cameras, the AVI video file produced can be imported into software packages like Registax 5/6 or Autostakkert 2 (AS!2). Both these freeware packages allow the selection of quality settings, stacking of multiple frames and wavelet sharpening as well as the application of final image enhancement. The software websites also contain some very useful tutorials for the novice.

The Solar Chat forum started by Stephen Ramsden to support his voluntary outreach program, Charlie Bates Solar Astronomy Project (see Appendix D), is a great repository of solar images obtained by amateurs around the world using different solar filters and cameras. The quality of the images presented daily clearly demonstrates not only the capabilities of the filters used by the members but also the high level of technical competence being achieved in the processing techniques. Specifications and performance for some of the most common fast frame camera are discussed in detail later in Chap. 6.

3.5   The Solar Surface: The Photosphere Features

As we will see, photospheric images can also be obtained using the digital spectro-heliograph (SHG) by selecting a wide bandwidth strip in a smooth continuum region of the recorded spectrum. The following sections provide more detail of the typical photospheric features which can be monitored and recorded in white light. Figure 3.7 shows well the typical high resolution features which can be recorded in white light.

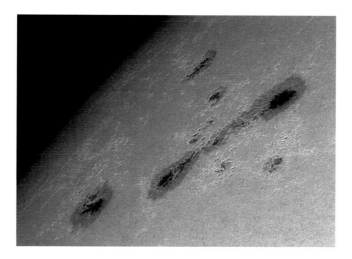

Fig. 3.7 The solar photosphere. Sunspot group AR2321 showing fibrils, light bridges background faculae and granulation (Bianconi)

3.5.1   Limb Darkening

When you look at a whole disk image of the Sun, the first thing you will notice is that the edge of the solar disk is much darker than the center. This is caused by the depth, approximately 200 km, of the photosphere atmosphere. At the top of the photosphere the temperature is only 4940°K, compared with 6370°K at the base. When we look at the center of the solar disk we see through the depth to the hotter base of the photosphere, whereas at the extreme solar limb we are looking at the cooler higher regions. This is well shown in Fig. 3.6a.

3.5.2   Granulation

Amateur telescopes with an aperture of 100 mm will start to show the solar surface covered with a mottled background. This was observed by Young and others as a coarse network of light and dark shaded cells or corrugations. These features are also referred to as Supergranulation. At higher resolution these cells are seen to be formed by collections of smaller individual granules. This rice grain-like pattern or granulation is made up of small cells approximately 1000 km across and separated by darker inter-granular lanes. The individual granules can show a wide variation in shape, brightness and size and are found throughout the solar surface and in the faculae.

It is thought that the granulation is the result of convection cells rising through the photosphere and may reach a height of 100 km above the photospheric surface. The average life of an individual granule is 18 min. A resolution of <1.5 arc second will record the individual cells. The cell structure is visible in Figs. 3.7 and 3.11.

3.5.3   Faculae

Bright ragged patches are also seen on the solar surface, usually more prominent towards the edge of the solar disk, where the limb darkening gives better contrast. These are faculae and are associated with active areas and sunspots. The faculae sit at a slightly higher level than the photosphere surface (see Fig. 3.7).

3.5.4   Active Areas and Sunspots

The visibility of active areas, sunspots and groups of sunspots follow the solar magnetic cycle, and peak on average every 11 years. They can be found between 5

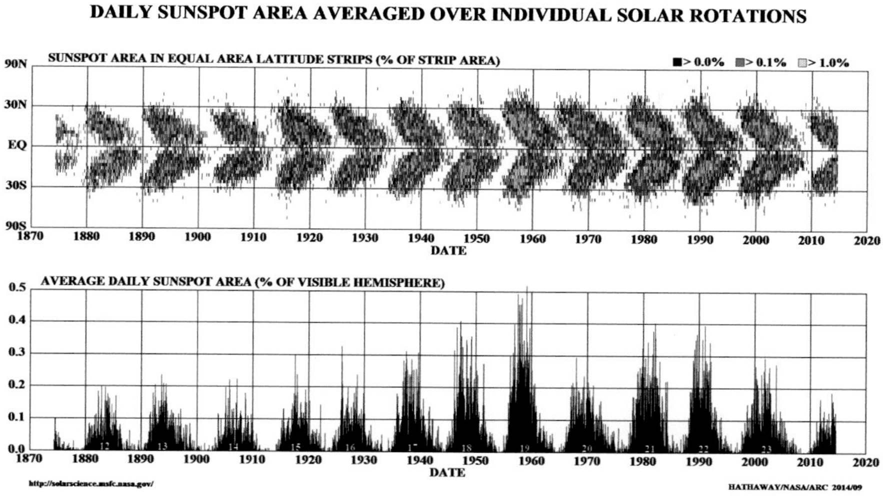

Fig. 3.8 Butterfly Diagram (NASA)

and 40° of solar latitude. Only very rarely are they on the solar equator or above 45°. The growth and life of a sunspot can vary from a few days to weeks. Additionally, larger sunspot groups can exist for more than a solar rotation. When the distribution of the sunspots is plotted over the solar cycle it clearly shows the transition from the higher latitudes to the lower latitudes as the cycle progresses. The resulting pattern (Fig. 3.8) gives rise to the well known "Butterfly Diagram."

Sunspots appear to grow from small dark pores and can develop into very large complex groups. The Zurich classification system is used to define the various stages of sunspot growth and decay (Fig. 3.9) All sunspots start and end as a type A group.

It is common to see bright facular areas develop into active areas — active region numbers and locations are regularly assigned by the National Oceanic and Atmospheric Administration Space Weather Prediction Center. This data is available on the Raben Systems, Inc. website. The typical interactive solar map is shown in Fig. 3.10.

Active areas are constantly monitored by the professionals as flares (X-ray flares and CME (Coronal Mass Ejections)) can quickly occur from these sites and cause disruption to electronic systems here on earth.

Umbra and Penumbra

As sunspots develop, their shape and structure become more complex. Starting from a small dark pore they generally become larger, displaying a dark area which

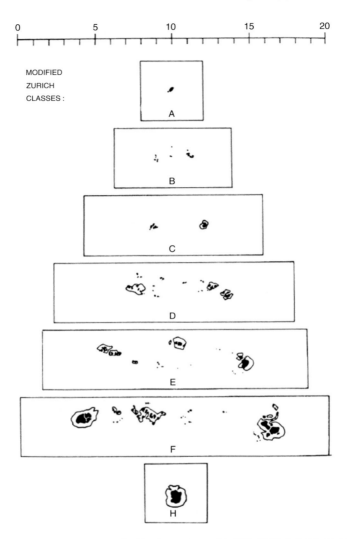

Fig. 3.9 Zurich Classes. Scale shown in degrees of longitude

eventually is seen surrounded by a slightly lighter area. The central dark area is called the Umbra, and the lighter surrounding area the Penumbra (see Fig. 3.11). These areas appear darker than the surrounding solar disk due to their lower temperature.

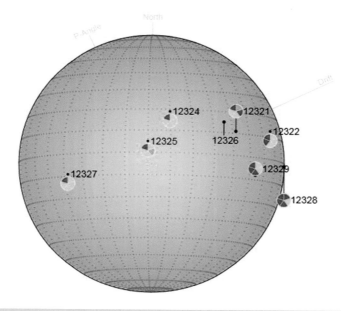

Fig. 3.10 Interactive solar map Bo −5.2 P −25.64 (Raben Systems, Inc.)

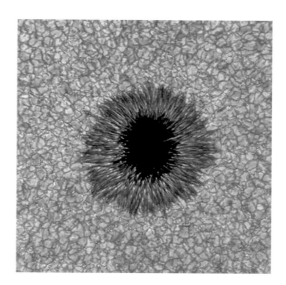

Fig. 3.11 Sunspot showing umbra and penumbra (BBSO)

Penumbral Filaments (Fibrils)

The magnetic flux surrounding the sunspots and active areas causes radial streaming of material, evident in the fine faculae/fibrils, threads or streamers that can be seen in the penumbra under good seeing. These are usually only 200 km wide. Similarly, faculae can form so-called light bridges across the dark umbra. These are clearly shown in Figs. 3.7 and 3.11.

Wilson Effect

It was noticed that when a regular sunspot approached the solar limb that the penumbra closest to the solar center appeared to reduce in width, more than the penumbra closer to the limb. Initially observed by the Scottish astronomer Alexander Wilson in 1769, this is opposite to the natural foreshortening effect expected. The results were consistent with the sunspot having a shallow dished profile. This profile therefore inferred that the central umbra lay at a lower level than the surrounding photospheric surface, and the penumbra was the inclined surface between them (see Fig. 3.12).

The change in apparent width of penumbra around a large circular sunspot can be recorded as the sunspot rotates across the solar disk and the amateur can record this effect for him/herself over a period of a few days. Later investigations have proved that the Wilson effect is probably caused by the difference in the light transmission of the umbra and penumbra.

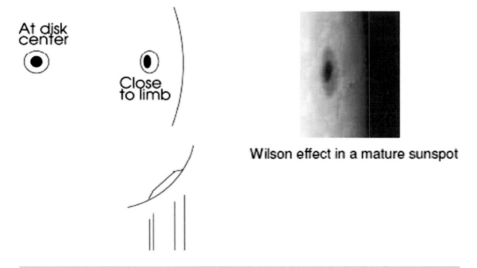

Wilson effect in a mature sunspot

Fig. 3.12 Wilson Effect

3.6   The Chromosphere: An Introduction

It was not until the spectroscope came into common usage with the amateur astronomers in the mid 1800s that the true nature of the red flames seen around the Sun during solar eclipses was determined. The outer reaches of the solar atmosphere, named the Chromosphere by Sharpey, was found to be the source of many spectral emission lines dominated by those of hydrogen, sodium and helium.

The brightest visible line, that of Hα, was carefully studied and the structures of the prominences faithfully recorded. These observations allowed dedicated observers like Secchi, Young, and others to determine the heights of the prominences and roughly measure the velocities of the material rising and falling within them. The possible cause of the prominences was assumed to be associated with sunspot activity, but as they could not observe them anywhere other than at the solar limb the correlation was restricted to matching their visibility to sunspot cycle.

It was however noted when comparing the distribution of the sunspot latitudes and daily numbers during the years 1853–1861 recorded by Carrington with the distribution of prominences by Secchi in 1871 that the sunspots were all contained within ±40° of latitude. The prominences by contrast were recorded all the way to the extreme polar regions. The prominences it seemed were not exactly following the sunspot distribution and to a degree they were independent of them. The favored explanation for where these prominences came from was "a quiet outpouring of heated hydrogen…an outrush issuing through the smaller pores of the solar surface, which abound near the poles as well as elsewhere" (Young).

Young was later amazed to observe the following on the 13th October 1880: "a small bright cloud appeared on that day…at an elevation of some 2 1/2′ (67,500 miles) above the limb, without any evident cause or any visible connection with the chromosphere below. It grew rapidly without any sensible rising or falling and in an hour developed into a large stratiform cloud, irregular on the upper surface, but nearly flat beneath. From this lower surface pendent filaments grew out." He thought this cloud-like development to be caused by condensation of the hydrogen gas due to local cooling or change in pressure.

Until the late 1890s, astronomers were restricted to observing the chromosphere and limited to interpreting its nature through the changes seen in the prominences. Many conflicting theories were debated as to what could be causing these impressive hydrogen clouds but there was no conclusive answer to be found.

Hale was among the new generation of solar observers in the latter part of the nineteenth century who tried to understand the nature of the chromosphere and in 1892 he re-visited Janssen's original idea of using a vibrating slit mounted on the spectroscope to allow the limb prominences to be better observed. Aided by the availability of sensitive photographic material he thought he could now move from visually recording the prominences to actually obtaining a lasting image of them. The early results obtained by Hale's first spectroheliograph surprised the astronomical community. He obtained the very first images of the chromosphere, in the light of calcium, at the CaK wavelength (3934 Å) and also in hydrogen with the Hβ line (4861 Å). Not only did he successfully record the available prominences he could, for the very first time show the chromospheric surface of the Sun. The new features which were found throughout the height of the chromosphere were given new names such as network flocculi, plage, spicules, dark filaments, fibrils etc. to compliment the prominences. A new landscape had been discovered and names given the prominent features. Subsequent spectroheliograph (SHG) discoveries over the following years of the solar magnetic field and the strong magnetic influences in and around the active areas and prominences quickly gave rise to the new field of solar magnetic studies.

The amateur today using narrowband solar filters and the digital spectroheliograph (SHG) can record the chromospheric features in more detail than Hale could ever achieve back in 1890 and some of the recent images posted on the Solar Chat forum by talented amateurs around the world are of excellent resolution and quality, showing solar detail which the professionals working on the study of the Sun a few years ago would envy. Technical abilities have improved vastly and citizen scientists are well-poised to take advantage.

3.7   Imaging the Chromosphere

In the remaining sections of this chapter we will review the specialized filters available and the results which can be obtained. We can be assured as further new technology becomes more available to the amateur that they will continue to be at the forefront of solar imaging. As we have seen, the very first images of the chromosphere were obtained by Hale using his newly invented spectroheliograph, a combination of telescope, spectrograph and moving slit plates in 1892. Similar instruments, updated for current technology in optics, diffraction gratings and CCD sensors are still in professional use today. After a hundred years the basic concept still works well, to the point that the digital spectroheliograph (SHG) as used by amateurs today would still be recognized by Hale.

The significant difference is that today we only need to use one entrance slit and use the software to produce a "virtual" second slit which isolates the target bandwidth to make the spectroheliogram image (see Chap. 4). This is easily constructed and can make use of readily available optics and fast frame cameras. By separating the solar light into a spectrum, with a suitable dispersion, the spectroheliograph (SHG) allows the selection of any wavelength and bandwidth resolution matching the best commercial filters (0.3 Å) to be recorded by the camera and extracted to form a solar image. The ability to select a wavelength gives the tremendous potential of being able to record features in CaK, Hα and many other interesting wavelengths using the same equipment and only one or two exposures.

Besides admiring the visual attributes of the very satisfying detailed images produced by the digital SHG, there's also the capability of doing some real science. By analyzing the images in the red and blue wings of the stronger wavelengths, Doppler effect, the Evershed effect, Zeeman effect, and more, we can monitor and record magnetic events and solar movements with much more accuracy than commercial filters. This will be discussed in detail in Chap. 5. The images presented in Chaps. 9 and 10 fully demonstrate the SHG capabilities.

Notwithstanding the above positive comments, there is one downside to the spectroheliograph (SHG)—its physical size. Even the latest generation of the digital version still needs long focal length optics to achieve the necessary dispersion for sub-angstrom resolution.

The sizes of the early spectroheliograph (SHG) soon put structural demands on the telescopes used. This in turn led to the development of the large dedicated solar tower telescopes like the 60 and 150 ft telescopes at Mt Wilson Observatory. Astronomers turned their attention to other options and started to consider the possibility of developing suitable filter elements which would allow imaging of the Sun in at least the critical wavelengths of CaK and Hα.

To obtain detailed imaging of the solar chromosphere, the filaments, active areas and the prominences requires a narrow band filter which only passes the light of the target wavelength. The wavelengths of Calcium, CaK (3934 Å) and Hydrogen, Hα (6563 Å) are the most common regions of interest, which mirrors the ongoing professional work in the same wavelengths. The range of commercial Hα (and CaK) filters available to the amateur is reviewed in the following sections. All are safe to use, and can provide suitable narrow bandwidth transmissions to allow the imaging of all the chromospheric features.

3.7.1 Imaging the Chromosphere: Hα Filters and Telescopes

Today there is a limited range of Hα filter systems available to the amateur. Due to the complex design of the filter elements necessary to give the narrow bandwidth required they are also much more expensive than conventional astronomical telescopes. A key parameter for chromosphere imaging is the bandwidth of the filter. As a minimum the bandwidth (FWHM) should be 2 Å or less, which allows the core area of the hydrogen line to be isolated and provides reasonable detail and contrast. Most commercial Hα filters target a better bandwidth of 1 Å or less and include an Energy Rejection Filter (ERF) element to protect the sensitive filter elements from excess heat load and damage.

The ERF is usually placed well to the front of the optical system, either at the front aperture or immediately in front of the narrow bandwidth filter elements. These ERF filters are usually dyed in mass colored glass filters (Sect. 2.1) equivalent to a Wratten 23A or 29 or Schott RG610/630, which suppresses all the ultraviolet (UV) and most of the visual wavelengths, but still passes all the infrared (IR) radiation. As an alternative, a suitable hard coated Induced Transmission Filter (ITF) can be used as an ERF. The Baader D-ERF is a classic example. The surface quality of λ/10 combined with the excellent multi-coated construction only allows a bandwidth of 300 Å centered on Hα to be transmitted, which then effectively suppresses all the UV, 90 % of the visual and all of the IR radiation.

The available Hα filters and associated telescopes are discussed below.

Thousand Oaks HAU System

The Thousand Oaks Optical Hα system is claimed to achieve 0.9 Å bandwidth. A red glass ERF is used at the telescope aperture followed by a multi-coated ITF filter assembly, the H-Alpha filter Unit (HAU) fitted to the rear of the telescope tube. The HAU can be tilted within the housing to provide central wavelength tuning. The optimum focal ratio required is f15 or greater which is achieved by varying the size of the ERF to suit the telescope focal length. For SCT's the ERF is mounted in an off center mask. Various sizes of ERF are available from 1.25″ to 6″ diameter. As we will see with some of the other Hα systems, tilting the filters is a common method of tuning the bandpass to the target central wavelength. This method does require that the design central wavelength of the filter be set just to the red side of Hα—tilting the filter element always moves the central wavelength toward the blue side of the spectrum. The field angle effects associated with multi- coated filters really restrict them to higher f ratios. It would be better to operate this filter around f30.

The Thousand Oaks system at least allows the amateur to make use of his/her existing telescopes (Barlow or TV Powermate lenses can be used to increase the f ratio if needed).

DayStar Filters

DayStar has been producing a wide range of narrowband solar filters for many years based on the mica spaced etalon concept developed by Dale Woods in 1970. All the Daystar premium filters are based on a mica solid crystal variation of the Fabry–Perot (F–P) filter design (Sect. 2.6.2) and require thermal control to maintain a target wavelength. All DayStar filter assemblies are rear mounted on the telescope OTA and require ERF protection when used with telescopes above 60 mm aperture. The design f ratio requirement for all the DayStar filters is f30 other than the Quark filters (see later) which can accept an f15 beam. DayStar now supplies complete range of solar telescopes which include their filters.

The Quantum filter is the DayStar premier flagship filter assembly. This filter assembly has a 32 mm clear diameter and designed for f30 systems. A sophisticated temperature controller (which can be interfaced to a PC) allows a central wavelength shift of ±1 Å with an accuracy of 0.1 Å. The Quantum is available in various bandwidths, 0.8/0.7/0.6/0.4/0.3 Å and offers some unusual CWL options—besides the Hα wavelength they also have filter versions for Na (D1 and D2), He (D3), and Hβ. 2A bandwidth version of the Quantum is available for CaH and CaK wavelengths.

The Quantum Ha filter appears to be the filter built into the Daystar solar telescopes—the 100 mm Skylight f32 and the SR-127 (Istar objective) 127 mm f32 telescope.

The ION Hα filter product range from DayStar now replaces the older tilt T-scanner filter and has a basic (non-computerized) heating/cooling temperature control. The clear aperture is less than the Quantum at 20 mm and an f30 system is still required. DayStar recommends adding a suitable ERF to telescopes with apertures greater than 60 mm when using the ION filter. Like the Quantum the ION is available in various bandwidths from 0.8 down to 0.3 Å.

The newcomer to the DayStar filter range is the Quark Hα filter. This utilizes the same F–P solid mica crystal etalon filter design and has a built-in ×4.3 telecentric Barlow. It is fitted with a similar basic temperature controller (but heating only) to the ION. With the added advantage of the built-in telecentric the Quark can be used on most telescopes with f ratios from f4 upwards. A minimal UV-IR filter as an ERF protection is recommended for telescope apertures above 80 mm. There are two versions marketed, the chromosphere model and the Prominence model and although DayStar does not quote specific bandwidth figures for the Quark, users claim the chromosphere model is around 0.5 Å bandwidth and up to 0.8 Å for the prominence version.

Solar Spectrum

Solar Spectrum also manufactures and supplies a range of mica etalons, very similar to DayStar. These Solar Observer series filters are designed for Hα wavelengths and are distributed in Europe through Baader Planetarium. The design f ratio is f25 or greater and the use of a front aperture ERF is recommended for telescopes above 60 mm. The Baader telecentric Barlow TZ2 (×2) or TZ4 (×4) are also recommended to increase the f ratio on faster refractor systems. Solar Spectrum appears to use a Thermoelectric controlled (TEC) oven with a set-point controller for all filters in the product range.

The base Solar Observer Series 1 has a clear etalon aperture of 19 mm and design bandwidths of 0.65/0.5/<0.3 Å. The intermediate Solar Observer Series 1.5 comes with a 25 mm aperture and similar bandwidth options. The Solar Observer Advanced filter is the largest etalon filter, fitted with a 32 mm clear diameter etalon and 0.8/0.65/0.5/<0.3 Å bandwidths.

The top of the Solar Spectrum range is the Solar Observer Research grade filters, similar to the Advanced series but with tighter design bandwidths of 0.5/0.3/0.2 Å.

Solar Scope

SolarScope, based in the Isle of Man, are renowned for the excellent quality of their solar filter products. The product range, designed for use at Hα wavelengths, is based on external air spaced etalon filters which can be mounted, using an adaptor cell, to the front of any refracting telescope. The etalons are all tilt tuned (Sect. 2.6.2) and the required blocking filter assembly, which fits at the rear of the OTA, is included.

All the SolarScope etalons are fitted with a front red glass ERF for safety and are designed for a bandwidth of 0.7 Å.

The stand alone front etalons systems (SF series) are available in four sizes, 50, 60, 70 and 100 mm clear diameter.

SolarScope also supply complete dedicated solar telescopes, the SolarView (SV) refractors. There are two models available the SV50 and SV60. The SV 50 comes fitted with an SF50 front etalon and has a focal length of 400 mm (f8). Similarly the SV60 has the larger SF60 front filter and a focal length of 480 mm (f8).

The SF series filters, like other front etalon systems, are suitable for double stacking. The bandpass achieved then reduces to 0.5 Å.

Coronado (Meade)

Coronado, later bought out by Meade Instruments in 2004, was established by David Lunt and started to supply amateur solar filters manufactured in the Isle of Man in the late 1990s; by early 2002 manufacturing was established in Tucson, Arizona. Coronado released in the same year, the first 40 mm aperture version of the Hα SolarMax tilt tuned (Sect. 2.6.2) air-spaced etalon, which was quickly followed (2003) by the visual 40 mm Personal Solar Telescope (PST) using for the first time a smaller (20 mm) internal air spaced tilt tuned etalon assembly.

Other larger models of the Solarmax external filters were added to the product line and in 2010 a modified SolarMax II (which included the Coronado RichView Tuning system) was introduced. The Coronado RichView tuning system incorporates a central pressure pad into the ERF assembly which, when rotated, applies pressure to the etalon and reduces the effective gap, giving the wavelength tuning. There have been mixed reviews from the users on the effectiveness of the RichView system.

The introduction of a CaK SolarMax70 and CaK PST version in 2005 was short lived and are no longer available.

Currently Coronado supply a complete range of tilt and RichView tuned air spaced etalon filters and dedicated solar telescopes.

Starting with the PST, this was designed as a visual Hα solar telescope with a bandwidth of 0.7 Å. The 40 mm f10 optical system combined with a 5 mm diameter blocking filter met the basic requirements of the visual observer but the restrictive backfocus spacing at the eyepiece holder makes adding a camera very difficult. Some users have had success by adding a Barlow lens at the bottom of the eyepiece holder.

The SolarMax Hα filter range covers 40/60/90 mm clear aperture RichView/tilt tuned using the supplied Tmax tuner. These filters can be mounted on any refractor using a suitable adaptor plate or used as a double stack on one of the Coronado SolarMax telescopes. Each filter has a red ERF built-in and a bandwidth of 0.7 Å.

Coronado also supplies their range of SolarMaxII dedicated Hα solar telescopes. These are available with apertures of 40/60 and 90 mm and all utilize a smaller internal tilt tuned etalon assembly. To complete the Hα filter system, blocking filters of 5/10/15/30 mm diameter are available to suit various telescope focal lengths (see Sect. 2.6.2).

Lunt Systems

David Lunt died in 2005 and his son Andy Lunt established Lunt Solar Systems in Tucson in 2008. The current product range includes a limited number of Hα external tilt tuned filters (50/60/100 mm) and a series of dedicated solar telescopes as well as a selection of blocking filters.

The 50 mm aperture Hα front etalon (tilt tuned) filter is available in two models—the LS50FHa which can be fitted to any refractor and a special LS50C which is designed as a double stack filter for the Lunt 50 solar telescope. The 60 mm filter, the LS60FHa is similar to the 50 mm LS50C but is said to be suitable for all refractors. At the top of the external filter range is the 100 mm version, the LS100FHa. All these filters are 0.7 Å bandwidth and require special adaptors to suit either the DS arrangement or the refractor being used.

To complement the above filters a range of blocking filters, suitable for various focal length instruments, Lunt can provide its BF600 (6 mm clear aperture), BF1200 (12 mm), BF1800 (18 mm) fitted to 1.25″ or 2″ diagonals or the larger BF3400 (34 mm) in a 2″ straight through version.

The Lunt dedicated telescopes (50/60/80/100/152 mm aperture) can all be supplied fitted with the internal pressure tuned etalon (see Sect. 2.6.2) option and are designed for <0.75 Å bandwidth.

The bottom of the Hα telescope range is the basic LS50THa (pressure tuned) 50 mm f7 aperture, followed by the LS60THa (60 mm f8.3) which is available in both the tilt tuned and pressure tuning version.

A larger telescope the 80 mm f7 LS80THa (pressure tuned) can be fitted with a second pressure tuned double stack module the LS80THa/PT/DSII, which reduces the effective bandwidth to 0.5 Å.

The 100 mm f7 version, the LS100THa is fitted with an internal <0.75 Å bandwidth pressure tuned etalon and can be fitted with the DSII/LS100THa double stack module to reduce the bandwidth.

The top of the Lunt range and the largest dedicated Ha telescope available to the amateur is the LS152THa. This is a 152 mm f6 telescope fitted with an internal 0.65 Å pressure tuned etalon. The double stack pressure tuned internal etalon assembly—DSII/LS152T is available to bring the bandwidth down to 0.5 Å.

Lunt systems are also the only current supplier offering narrowband CaK filters.

Any of the commercial narrowband solar filters detailed can be safely used for chromospheric imaging.

Note that each of these filters is pre-tuned to the target central (CWL) wavelength, usually allowing only a limited range to be selected. A Hα filter can't be used for CaK observing and vice versa. This means that if the amateur wants to observe and record the Sun in different wavelengths of light he/she has to invest in multiple filters.

3.7.2 CaK Filters

There are limited CaK filter options available to the amateur. DayStar and Solar Spectrum offer CaK (and other wavelength filters) on special demand only and don't appear to normally hold stock.

Baader supply a multi-coated CaK filter with an 80 Å wide bandwidth, centered at 3950 Å. This bandwidth provides ample contrast for high-resolution imaging of super granulation, flares, and other features that are prominent in CaK but certainly not the detail to be imaged in the K2 region of the CaK line (see later).

Lunt Systems has a selection of CaK filter modules available in its product range the B600/B1200/B1800 and the B3400. These rear mounted multicoated filters can be used on any refracting telescope (Lunt recommend <100 mm aperture) and have a bandwidth of 2.4 Å, centered on 3934 Å.

Imaging the Chromosphere: Cameras

The cameras detailed in Sects. 3.5 and 6.7 can also be applied to chromosphere narrow band imaging. The same general processing techniques apply. There is however the significant issue that narrowband filters and CCD chips don't always play well together! Imaging in extreme narrowband monochromatic wavelengths can cause the appearance of Newton Rings. These are seen as a light and dark series of straight (or slightly curved) bands extending across the image and are caused mainly by interference patterns generated in the very narrow gap between the thin glass protection plate and the underlying surface of the actual CCD chip. Fringing is a more subtle form of interference pattern and is visible as light and dark distortions spreading throughout the image.

Howell in his book on astronomical CCD's explains that fringing is caused by interference within the CCD or by light which passes through the CCD array and subsequently reflects back into the array. As white light imaging usually entails using much broader wavelength coverage, they generally do not appear. Newton Rings don't appear on all cameras, some seem to be more prone to them than others—i.e. the ZWO ASI120MM. It has also been noted that using longer effective focal lengths, when imaging detail in and around sunspots for instance, seems to aggravate the interference. As it's impossible to change the glass cover plate on the CCD chip various remedies have been tried by the amateur. The most successful method is to introduce a slight tilt to the camera which can sometimes reduce, and in the best cases, remove the Newton Rings. The tilting of the camera relative to the optical axis increases slightly the effective width of the gap thereby reducing the interference. The fallback is to prepare flats (Sect. 8.4) and apply them to minimize the Newton Rings. And if it is any consolation even the professionals have problems with Newton rings and fringing on their CCD cameras.

3.7.3   Digital Spectroheliograph (SHG)

The digital spectroheliograph (SHG), which will be discussed in detail in a later chapter can be used to effectively select and extract a very narrow bandwidth region of the spectrum, either in the continuum parts of the solar spectrum, to show the surface photospheric features or in the core or the wings of one of the prominent absorption lines and then use these strips to reconstruct an image to show all of the chromospheric features. By recording a high dispersion solar spectral video AVI the available software allows the extraction of bandwidths down to 0.2 Å centered on any selected central wavelength (CWL). This flexibility allows the digital SHG user to process spectroheliograms which will show all the image detail normally associated with the more expensive commercial filters. As can be seen from the examples in Figs. 3.13a, b and 3.14a, b the quality of the digital SHG results being obtained with the current instruments are very similar to the image detail achieved with the commercial filters. Other excellent examples of the images being obtained by the amateur and their SHG instruments are shown in Chap. 8. It should also be noted that the SHG has the potential to provide unique data which allows us to record and examine many of the important magnetic features of the Sun as illustrated in Chap. 9.

a b

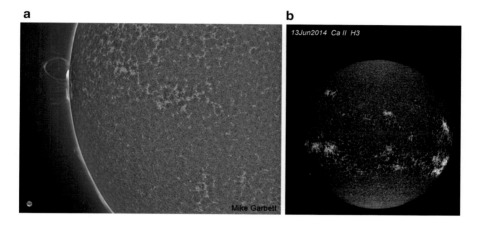

Fig. 3.13 Wilson Effect. (a) Lunt CaK image (M Garbett). (b) SHG CaK image (P Zetner)

a b

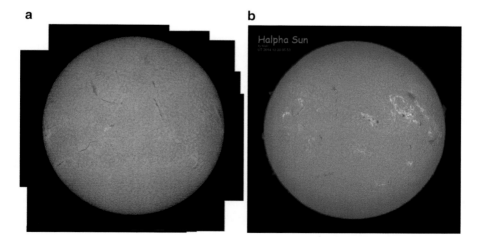

Fig. 3.14 (a) Hα mosaic (PST) (J Bennett). (b) Hα SHG (Wah)

3.7.4 Chromosphere Imaging Terminology

On Band, Off Band, Core and Wings

The terminology used in chromospheric observations needs a little explaining. You will come across words like "On/Off band" and "red and blue wings" which are not used in any other context. All the absorption features recorded in the solar spectrum, whether by using commercial Hα and CaK filters, or the digital SHG are the result of the complex interaction of the atoms with the surrounding solar conditions (temperature/pressure/magnetic fields etc.) which in turn generates varying absorption line widths and structures.

The profiles shown in Fig. 3.15 map the intensity of the absorption line against wavelength. Each element line has a particular shape which is a result of the amount of that material in the solar mass, its energy level, and its temperature. As a result of these factors we see a line shape which shows a deep narrow core the bottom of which can have an intensity, in the case of Hα, of only 16 % of the surrounding continuum. This core for Hα, is 2 Å wide. Superimposed on this core are the wings, which extend the top section of the core into the continuum on either side. The shorter wavelength gives the blue wing its appearance and the longer wavelengths the red wing. The depth and hence the intensity of the wings can vary, but they are never as intense as the core. The Hα wings can extend up to 8 Å on either side of the core.

When the observed wavelength coincides with the central wavelength of the core we are said to be observing on band—this is where the resulting image reaches maximum contrast due to the deep core structure.

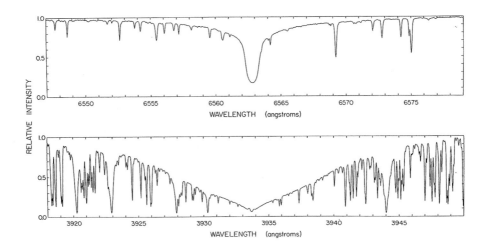

Fig. 3.15 Comparison of the Hα and the CaK absorption lines (Noyes)

If the filter or the selected spectroheliograph (SHG) wavelength is tuned such that the central wavelength is no longer in/on the core, it is said to be off band. In this case we are starting to observe in the transition zone from the core to the wings. When the selected target wavelength is beyond 0.5 Å of the core then we start to observe in the red/blue wing dependent on direction. As we move further and further from the core, we can end up with a loss of image contrast and detail as the surrounding continuum starts to contribute parasitic light to our narrowband image.

To record accurately the differences between the core and wing details in our image, the bandwidth of the filter system must be suitable for the absorption line under observation. Most of the comparison professional images we see were taken within ±1 Å of the central core wavelength at bandwidths of 0.5 Å, or better.

A 2 Å bandwidth, centered on the Hα will record all the core detail (a good solution when imaging the prominences), but will not give the same level of detail/contrast that a 0.5 Å bandwidth, which would isolate the deeper core structure.

Obviously the narrower the bandwidth used, the further we can record into the wings without continuum leakage.

The Importance of the Core and the Wings

Early chromospheric observers like Hale and others were surprised to find that the solar disk looked completely different at different wavelengths (CaK and Hβ), not only that, the image could dramatically change from the core to the wing wavelengths. Figure 3.16 details the various solar features visible at different target wavelengths.

The reason for the differences is related to the varying energy conditions and heights of the elements within the chromosphere. This will be discussed in more detail in Sect. 5.3. We can therefore use the differing images obtained in the core and the wings to determine a chromospheric profile as shown in Figs. 3.2 and 3.16.

To recap: the absorption line of Hα comprises a deep dark core with extended wings spreading out on either side. These wings are greater than 10 Å wide as they reach the surrounding continuum. The CaK line appears much wider, shallower, with wings extending over 20 Å and the core less pronounced as shown in Fig. 3.15, but it does show some additional and very important features close to the center. The whole CaK line is called the K1 line, with K1v and K1r being used to describe the extended blue (violet) and red wings. Within 1 Å of the core (K3) the K2 regions (and the K2v and K2r wings) are used to record the full range of CaK features.

Figure 3.17, shows the variation to be found around the K2/K3 region; A represents the profile found in a bright facular region, B similar but near it's edge, C bright flocculi and D the quiet cell areas. The chromosphere in Hα shows a general mottled appearance; the small clouds, both light and dark, are called hydrogen flocculi and are reminiscent of the photospheric granulation. These flocculation features extend high into the chromosphere and show well in the CaK wavelength. The K1 region shows the lower flocculi and the K2 components the higher altitude formations.

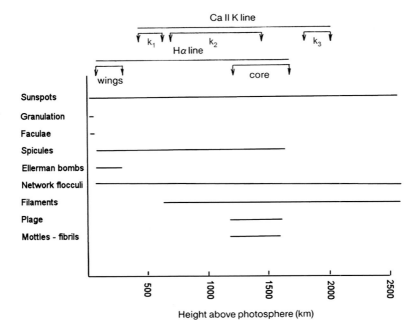

Fig. 3.16 Solar features recorded at CaK and Hα wavelengths

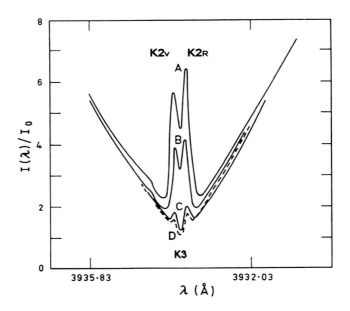

Fig. 3.17 K2 and K3 regions of the CaK line (Bray and Loughhead)

3.8 The Solar Atmosphere: Chromospheric Features

Hα (6563 Å) and CaK (3934 Å) wavelengths have long been used for observations of the chromosphere. The professionals differentiate between the quiet Sun features and those peculiar to the active Sun, recorded in and around active solar areas and sunspots. A collection of images showing the significant chromosphere features in Hα and CaK wavelengths are shown in Fig. 3.18.

3.8.1 Spicules

Spicules are extended, narrow column-like features less than 800 km (<1 arc second) wide by 10,000 km (>3 arc seconds) high and exist for an average life of less than 10 min. They cover the whole solar disk but are most prominent when recorded at the limb.

They are visible at the central core of Hα, but best seen in the blue/red wings ±0.5 Å away from the center wavelength (see the great example in Zirin's "Astrophysics of the sun", p. 161). A 1 Å bandwidth will show the spicule "bulk" (my words—hairy edge) and a 0.2 Å or narrower bandwidth will show the individual spicule motion. Observing and recording spicules seems to rely more on aperture/seeing to give good resolution to record the 3 arc seconds height above the photosphere than they rely on bandwidth (see Fig. 3.18).

3.8.2 Mottles and Fibrils

The solar surface when viewed in Hα seems to be covered by a mottled blanket of small features, the hydrogen flocculi. The spicules are not evident near the solar center due to the reduced contrast with the brighter surface. Under high magnification and good seeing conditions the edges of the mottles can be resolved to short lived fibrils, small dark elongated features, a bit reminiscent of long grass swaying in wind. Where there is an active area the mottles and fibrils show strong alignment with the magnetic field and appear to be distorted close to filaments.

3.8.3 Plage

In and around active areas, with or without established sunspots, small bright patches are seen in the central Hα wavelength. These bright, relatively featureless areas, are known as plage (French: Plage=Beach) (Fig. 3.18). The surface fibrils/mottles can be seen to have a rough radial alignment with the plage areas. This is thought to be caused by the higher magnetic field existing in active areas around the plage regions.

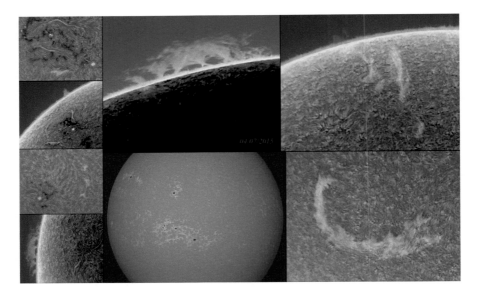

3.8.4   Filaments and Prominences

A very noticeable feature of the chromosphere is the long dark wispy elongated cloud which stands out against the brighter surface. When seen on the disk these clouds are called filaments, and when projected from the solar limb called prominences. A filament and a prominence are therefore the same thing. When a filament can be traced from the disk to beyond the edge it is called a filaprom (see Fig. 3.18).

These are visible in both Hα and CaK and follow the solar cycle. More sunspots visible mean more filaments.

The filament is usually seen as an arch between one sunspot and another active area due to the opposing magnetic polarity. The movement of the filaments upwards, across and downward can be traced by working off-band to register the associated Doppler shifts.

Hedgerow Prominence

The hedgerow prominence is the most common type of prominence. It is associated with the quiet regions of the Sun i.e. away from active areas and sunspots. The hedgerow prominences, as the name suggests show separated vertical linear features topped by an extended variegated cloud and are also known as quiescent prominences.

Some of these hedgerow prominences can grow to enormous sizes (up to 600,000 km long and heights of 10,000 km) and can show intricate lacy patterns as the material slowly falls back towards the surface as shown in Fig. 3.19. With careful Doppler measurements, the movements of the material can be clearly mapped and recorded. With an average lifetime of over 4 weeks, some can last more than one solar rotation.

Flares, Surges/Spray Prominences

Flares and surges are associated with active areas where the magnetic field controls the shape and movement of the material. These are normally short lived and very dynamic. The moustaches (described at length later) associated with flares indicate they at least start at low altitude, just in and above the photosphere, and extend into the chromosphere. Surges and smaller flares when visible close to the solar limb can give rise to short lived but spectacular spray prominences.

Flares, besides causing material ejections into the solar wind and possible magnetic storms here on Earth, with the associated failures of electronic equipment, also result in the rapid expulsion of material which can be seen as bright flames and sprays in Hα.

The GEOS satellite records solar flares in X-ray which are classified by their total power. An A Class flare being the least powerful and X Class the greatest.

Fig. 3.19 Quiescent or Hedgerow prominence (Friedman)

Loops

Loops are prominence like features which dramatically follow the magnetic lines forming multi-strand loops. An unusual CaK loop was captured by Klepp in July 2015 as shown in Fig. 3.13a.

Coronal Rain

Coronal rain occurs when plasma ejected into the solar atmosphere cools and gets attracted downwards by magnetic field lines on the Sun's surface. It can appear as a cascading waterfall below larger quiescent prominences. Figure 3.19 shows this effect very well.

3.8.5   The Chromospheric Network

Overlying the whole solar surface, visible as a network of cells (most noticeably in the CaK wavelength) is a mat of material which was given the name as "flocculi" by Hale. These flocculi elements appear to form larger complex cells of varying shapes and sizes. This is the chromospheric network. Brighter flocculi areas are seen around active areas. These supergranulation cells are situated high in the chromosphere and can exceed 30,000 km in size and have a life of 10 h or more (see Fig. 3.13b).

3.8.6   Ellerman Bombs (Moustaches)

Around active areas and sunspots the magnetic field causes small elongated arches, or fibrils. These always show upward movement at the middle and downward flow at the ends. In the wings of Ha +0.7 Å will only show the ends of the fibril, and −0.7 Å the top. When the magnetic field is broken or quickly collapses in one of these small fibrils, the resulting "flash" is recorded as a bright point which can last for many minutes (10–40 min). This was first recorded by Ellerman who said the looked like the flash of a bomb going off, hence the current "Ellerman bombs" description. A more realistic description was given by Severny—"Moustaches", based on the profile of the extended broad emission wings seen in the spectrum which resembled a moustache with a gap in the middle. The emissions associated with moustaches are never seen on-band, hence the gap.

Figure 3.20 clearly shows the bright Ellerman bombs clustered around the sunspot (see Sect. 5.5 and Chap. 10 for further detail).

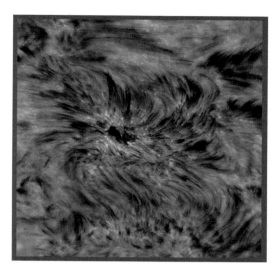

Fig. 3.20 Active area recorded in Hα (*Blue wing*) on 14 June 2010 (BBSO)

3.8.7   Morton Wave

The Morton wave is a dynamic chromospheric feature which traces the edge of an expanding shock wave. The source is generally a large flare (Type II burst), and the wave moves outwards at constant speed (approximately 1000 km/s) until it reaches a magnetic boundary (usually a filament or active area).

The front of the wave can be seen in red/blue shifted Hα with a red edge (downward motion) followed by the blue edge (upward motion).

Further Reading

Secchi, P.A.: Le Soleil, 2 Vols, Gauthier-Villars (1875, 1877)
Young, C.A.: The Sun. Kegan Paul Trench&Co (1882)
Bray, R.J., Loughhead, R.E.: Sunspots, Dover (1979)
Bray, R.J., Loughhead, R.E.: The Solar Chromosphere, Chapman and Hall (1974)
Bray, R.J., Loughhead, R.E. and Durrant, C.J.: The Solar Granulation, Chapman and Hall (1985)
Phillips, K.J.H.: Guide to the Sun, CUP (1992)
Zirin, H.: Astrophysics of the Sun, CUP (1988)
Abetti, G.: The Sun, Faber and Faber (1963)
Baxter, W.M.: The Sun and the Amateur Astronomer, Lutterworth Press (1963)
Noyes, R.W.: The Sun - our star, Harvard University Press (1982)
Jager, C. and Svestka, Z. (Eds.): Progress in Solar Physics : Review Papers Invited to Celebrate the Centennial Volume of Solar Physics, Springer (1996)
Howell, S.B: Handbook of CCD Astronomy, CUP (2010)

Webpages

http://iopscience.iop.org/1742-6596/440/1/012007/pdf/1742-6596_440_1_012007.pdf
http://adsabs.harvard.edu/full/1966SSRv....5..388S
http://en.wikipedia.org/wiki/Solar_flare
http://bass2000.obspm.fr/solar_spect.php?step=1
http://fermi.jhuapl.edu/liege/s04_0000.html
http://www.petermeadows.com/html/stonyhurst.html
http://www.atoptics.co.uk/tiltgrid.htm
Space Weather Prediction Center
https://www.raben.com/maps/
http://gong.nso.edu/
http://www.iiap.res.in/ever/PDF/bagare.pdf
http://sohowww.nascom.nasa.gov/data/realtime/
http://www.astronomie.be/registax/
http://www.autostakkert.com/
http://www.meade.com/products/coronado.html
http://luntsolarsystems.com/
http://www.solarscope.co.uk/sv-range.html
http://daystarfilters.com/Telescope.shtml
http://www.skyandtelescope.com/observing/celestial-objects-to-watch/solar-filter-safety/
http://www.company7.com/baader/options/asolar.html
http://www.company7.com/baader/options/solarcontinuum.html
http://www.physics.rutgers.edu/ugrad/387/Mulligan98.pdf
http://www.thousandoaksoptical.com/halpha.html
https://www.daystarfilters.com/ION/ION.shtml
http://www.daystarfilters.com/downloads/QuarkFlyer.pdf
http://www.solarscope.co.uk/
http://www.baader-planetarium.com/pdf/solarspectrum_e.pdf
http://www.meade.com/products/coronado.html

Chapter 4

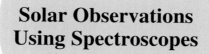

Solar Observations Using Spectroscopes

Since the time of Isaac Newton (1642–1727) scientists and astronomers have been trying to come to terms with the fact that all lenses produce a spectrum. Newton's early experimentation with prisms incorrectly led him to conclude that a glass lens would always present this spectrum of light as a chromatic aberration and was therefore of limited value to astronomical telescopes. If nothing else, this encouraged him to work on reflective optics and the "Newtonian" telescope was born.

Newton's conclusions, though accepted by some, failed to discourage others such as optician William Dollond (1706–1761) from investigating a design combined of lenses constructed from different types of glass. It was thought that this could reduce the seemingly false color found in the simple lenses then in use. The challenge of producing a quality glass demanded an improvement in the technology used to view and understand the spectrum. This technology was not immediately forthcoming and it was only in the early 1800s that the basic spectroscope was developed. As instrumentation improved, so did the ability to resolve the spectrum to show the dark and bright lines associated with various chemicals. The analysis of these lines enabled a better understanding of the physical processes that caused these lines to be formed in the first place.

An often quoted and unfortunate comment was made by the French philosopher Auguste Compte (1798–1857) in reference to the general state of knowledge on the heavenly bodies: "We can never learn their internal constitution, nor, in regards to some of them, how heat is absorbed by their atmosphere." Compte was soon proved to have grossly underestimated the advances in science within his lifetime. The work of Gustav Kirchhoff (1824–1887) and Robert Bunsen (1811–1899) coupled

© Springer International Publishing Switzerland 2016
K.M. Harrison, *Imaging Sunlight Using a Digital Spectroheliograph*,
The Patrick Moore Practical Astronomy Series, DOI 10.1007/978-3-319-24874-5_4

with the ongoing development of the spectroscope did indeed give the astronomers the tools to see the "heavenly bodies" in a new light. The advent of the spectroscope was to go on to change astronomy for ever. The era of the Victorian gentlemen and their visual observations was about to be transformed into a scientific community based at professional observatories and dedicated to the pursuit of the new astronomy of astrophysics.

Making use of the new technologies of photography, photometry, and the spectroscope these scientists quickly commenced their investigations of the physical characteristics of the celestial bodies, including the luminosity temperature, the conditions of the solar atmosphere and the qualitative and quantitative composition of the Sun and stars. This chapter traces the early beginnings of astronomical spectroscopy and acknowledges the key players involved.

4.2 Early Spectroscopy

Isaac Newton used prisms in the mid 1600s to show that the "refrangibility"—what we'd now call the refraction/dispersion—of the prism caused incident light to be broken down into a spectrum. In one of his most famous experiments with light, he demonstrated that colored light dispersed by one prism when viewed through a second prism did not produce further spectra but an increase in the dispersion only. It was found in the early 1700s that the combination of two lenses, one made from flint glass and the other from crown glass and called an achromatic objective, corrected much of the chromatic aberration's false color. These new objective were soon to be seen in the hands of professional and amateur astronomer and until the revival of the reflector with the silver on glass coatings in 1857 by Jean Léon Foucault (1819–1868), the achromatic refractor became the preferred telescope in all new observatories.

Unfortunately the quality of this early optical glass was very poor and many trial batches of glass had to be made to find a suitable combination for color correction.

Josef von Fraunhofer (1787–1826), a young German optician in the early 1800s, was determined to measure the dispersion of the glass he was producing. By recording both the dispersion and the associated manufacturing processes he was able to achieve better control over the types and quality of glass he produced. His results produced the best telescope objective lenses of the time and his large telescopes were the envy of all astronomers.

In order to view and measure the dispersion of the glass being manufactured, Fraunhofer developed a measuring device, which he called a "spectroscope". Based on a small telescope on a rotating platform named a theodolite, the telescope objective was directed towards a sample glass 60° prism illuminated by a narrow slit in his window-blind at a distance of 20′–30′. The use of the slit gave a cleaner more detailed image of the spectrum. Fraunhofer also noticed that sunlight, when dispersed by his prisms, always seemed to be crossed by dark lines in the same positions and

Fig. 4.1 Fraunhofer's spectrum (NASA)

eventually he mapped some 324 lines; the more prominent ones are still called "Fraunhofer lines"(see Fig. 4.1). Although he didn't understand how these lines were produced, he made good use of them as "standard wavelengths" for his optical testing.

Some of these dark lines had been previously recorded by the English scientist William Wollaston (1766–1828) in 1802, when he illuminated a prism with sunlight via a narrow opening an "elongated crevice 1/20″ wide" at a distance of 10′–12′, but he associated the lines with the apparently natural boundaries of the spectral colors and did not pursue his investigations.

The basic spectroscope soon gained acceptance in the scientific community and was widely used in laboratories for early chemical spectral investigations. By 1839 a collimator and slit assembly was added to the basic theodolite telescope to further improve the resolution and provide a more compact and manageable instrument.

The first to adopt the use of the entrance slit and collimator in the spectrograph layout seems to have been the French scientist Jacques Babinet (1794–1872) in 1839. In 1840 the famous English optician William Simms (1793–1860) fully described the arrangement and for the first time used the word collimator to describe the front telescope. Both the viewing telescope and collimator were mounted on a rotating table fitted with a graduated circle which also supported the prism. This arrangement, for the first time, allowed the deviation angle of the various wavelengths of light to be accurately measured for various glass prisms.

In a continued effort to further improve the dispersion and resolution of the spectroscope, it was soon common practice to increase the number of prisms used. The spectroscope used by Gustav Kirchhoff for his solar investigations made use of three 45° and one 60° prisms.

4.3   Identification of Elements Using the Spectrum

4.3.1   Kirchhoff Laws

In 1859, the German chemist Bunsen and physicist Kirchhoff used a spectroscope to finally determine the cause and source of the absorption lines noted by Fraunhofer. They found that each element produces a unique fingerprint series of emission/absorption lines. The later Kirchhoff laws, listed below, were established on their findings:

1. An incandescent solid or gas under high pressure will produce a continuum spectrum.
2. An incandescent gas under low pressure will produce an emission spectrum.
3. A continuous spectrum viewed through a low density gas at low temperature will produce an absorption line spectrum.

This knowledge became the foundation of the new science of astrophysics. Over the next decades more and more elements were identified in the laboratory and also found in the solar spectrum. (Appendix A lists the more important lines in the visual solar spectrum.)

Measurement Units: Angstrom or nm?

The unit used for visible wavelength measurement was standardized in the mid 1800s by the work of the Swedish astronomer Angstrom, established at $1 \text{ Å} = 10^{-9}\text{m}$. On this scale the visible spectrum covers the range from 4000 Å (blue) to 7000 Å (red). By the 1960s the International System of Units (SI) redefined all the measures in terms of the meter and one of the recognized official measures of visible wavelengths became the nanometer (nm) which is 1^{-10} m.

Hence 1 nm = 10 Å.

Angstrom units are still widely used in the astronomical community.

4.4   Early Use of Spectroscopy in Solar Observing

Since the early 1800s the dramatic appearance of red flames visible at the edge of the moon during the total solar eclipses had attracted the attention of astronomers. Initially they were thought to be associated with the lunar atmosphere and it was only with the total eclipse observations and the first successful photographs in 1860 that confirmation was obtained that the red flames were actually prominences that formed part of the solar atmosphere. Norman Lockyer (1836–1920), the English physicist and astronomer, was the first to apply the spectroscope to a solar feature, as distinct from observing the spectrum from the whole solar disk. At this time there was still some debate as to whether sunspots were hotter or cooler than the surrounding surface. If they were cooler, then the spectrum should show absorption lines; if hotter, the lines should be in emission. Lockyer set up his spectroscope behind a screen in which he had cut a fine slit. He then projected the Sun's image from his telescope onto the screen, and positioned the slit across a sunspot. Lockyer found the Fraunhofer lines, rather than being in emission, were widened across the sunspot. This confirmed to Lockyer that sunspots were indeed cooler than the surrounding photosphere. His published paper on the subject concluded with the following prophetic comment: "May not the spectroscope afford us evidence of the existence of the "red flames" which total eclipses have revealed to us in the Sun's atmosphere; although they escape all other methods of observations at other times?"

The introduction of the spectroscope gave astronomers an opportunity to carry out further investigations at the subsequent total eclipse of August 1868 in India. The French astronomer Jules Janssen (1824–1907) and others there applied their spectroscopes to the issue of the red flames. There was also the question of the layer of smaller flames (called a "sierra" by George B. Airy (1801–1892), who had seen them at the 1851 total eclipse visible from Sweden) projecting just above the photosphere.

It was quickly apparent to all the spectroscopic observers that the red flames, or prominences, were the result of the emission line of hydrogen, Hα (6563 Å). At last they had definite proof of the nature of the solar atmosphere, named the chromosphere by Doctor William Sharpey (1802–1880). The chromosphere was a gaseous atmosphere surrounding the Sun and extending upwards from the solar surface by many thousand kilometers. This is the region where the prominences were recorded, a zone where the absorption line as seen in the usual solar spectrum were reversed and seen in emission.

Janssen was so impressed by the brilliance of the hydrogen line that he felt confident that he would be able to view it in his spectroscope without the need for a solar eclipse. The following morning he set up his equipment and positioned the entrance slit of the spectroscope on the solar limb at the same position at which he had recorded a very bright prominence the previous day. The emission line spectrum was found unchanged and he worked to identify the few other lines he saw. Once again the Hα line was the brightest but he was puzzled to see another bright line in the spectrum close to the known positions of the sodium D1/D2 lines, which seemed to be positioned more towards the blue side of the spectrum. This same line had been found independently by Lockyer.

The astronomers called it the D3 line (5376 Å) and associated it with the yet-unknown element Helium. It was not until 1895 that this element was positively confirmed when Helium was finally isolated in the laboratory by Sir William Ramsey (1852–1916).

Janssen found that by setting the spectroscope to the Hα wavelength and positioning the slit as he had previously, the previously observed prominence was still noticeable. He also found that by slightly moving the telescope around the edge of the Sun he could see different sections of the prominences. For the next few days Janssen enjoyed the luxury of being the first astronomer to witness the extended solar chromosphere.

In one of those twists of fate, unbeknownst to Janssen the English spectroscopist Lockyer had independently also reached the same conclusion: that the chromospheric light, if associated with a glowing gas, should show a bright line spectrum. These individual emission lines should be bright enough to be viewed through a suitable spectroscope. Following up on this idea, he commissioned a larger spectroscope to attempt the observations. Just as the instrument was finally commissioned he heard of Janssen's successes on the Indian eclipse and immediately started observing the edge of the Sun. Lockyer traced the bright emissions around the complete edge of the Sun and found some prominences. He concluded that the chromosphere totally covered the Sun and that the prominences were extensions of this solar atmosphere. In recognition of the achievements of Janssen and Lockyer, in 1872 the French government presented them with a gold medal showing both their effigies. Some 6 months after the Indian eclipse in February 1869, William Huggins (1824–1910) developed a technique to allow the observation of solar prominences without the need for a solar eclipse.

By setting his spectroscope slit tangentially to the solar disk, and tuning the spectroscope to the Hα wavelength, he could observe a section of any prominence active on the edge. Once located, the slit gap could be increased to allow the full extent of the prominence to be seen. Huggins' first recorded sketch of a prominence using this method is illustrated in Fig. 4.2. The contrast difference in the brightness of the prominence to the immediate background sky made observation relatively easy. This became the standard method of observing prominences for the next 20 years (see Sect. 6.7.11 for further details).

There was much discussion during the late 1800s on the possible nature and origin of the prominences but no new information was added to our knowledge of the chromosphere. Observing the solar photosphere continued to be restricted to the white light observation of granulation and sunspot activity. Occasionally the spectroscope would show some activity in and around sunspots but few details were accurately recorded.

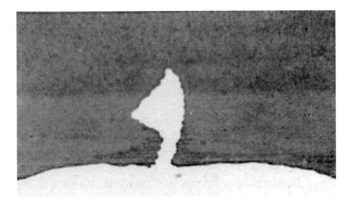

Fig. 4.2 Huggins' first recorded prominence (Huggins)

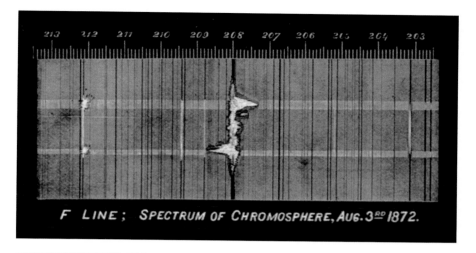

Fig. 4.3 Distortions in the Hβ line as observed by Young (Young)

Young, observing in 1872, recorded distortions in the Hβ line which indicated velocities of 230 miles/s or 385 km/s (see Fig. 4.3). He was at a loss to explain as to how such velocities could be achieved. It would take another 20 years to unravel the mystery.

It was only the work of George. E. Hale (1868–1938) and Henri Deslandres (1853–1948) in developing a functional spectroheliograph (SHG) in the early 1890s that allowed us finally to actually image the chromosphere across the whole of the solar surface and appreciate it in all its glory.

4.5   Spectrographs and Spectroheliographs (SHG)

As we have seen, the early spectroscopes were based on prisms and were purely visual instruments.

The development of the colloidal film in the late 1870s provided the first real opportunity to record and take photographs of spectra, albeit in limited wavelengths. It wasn't until the late 1880s that diffraction gratings (Sect. 6.6) became available. The first generation gratings, produced by the American physicist Henry Rowland (1848–1901) and others, were very expensive and only used by professional observatories. The advent of replica gratings in the early twentieth century made them more available to the amateur.

Even with the improved quality and dispersion of the gratings, applying photography to astronomical spectroscopy did not prove to be an easy task. Early photographic trials at John Hopkins University in the 1890s only achieved a usable spectrum of Sirius after a 40 min exposure, for instance. Obtaining photographs of the solar spectrum however was more a more straight forward affair.

In 1883 Henry Rowland at John Hopkins University used his recently invented concave reflection grating in combination with the latest photographic plates to produce a highly detailed solar spectrum with an accuracy of 0.01 Å.

Jules Janssen, Johann Zollner (1834–1882), Norman Lockyer and others had already suggested the possibility of using a vibrating slit to view the prominences, taking advantage of the eye's persistence of vision (see later). Unfortunately none of these early experimenters were able to implement a workable solution. It wasn't until 1890 that George E. Hale and Henri Deslandes independently invented the spectroheliograph. Hale's first successful SHG was completed in 1892 and installed on the 12″ (320 mm) Brashear refractor in his Kenwood observatory. Hale's original arrangement is well explained in his 1890 paper. In summary, it used a pair of slits 82 mm long to accommodate a solar image of 50 mm diameter with a 100 mm 568 l/mm Rowland reflection grating. Both the collimator and imaging lenses were 1080 mm focal length. His concept was to project the image of the Sun onto the entrance slit of the spectroscope. The entrance slit was connected to a second slit which sat immediately in front of the photographic plate. This second slit was aligned with the spectrum produced by the spectroscope and centered on the wavelength under observation, thereby only allowing a very narrow slice of light to reach the photographic plate (see Fig. 4.4). Today we would call it a monochromator.

As the telescope scanned across the Sun using the mount's declination drive, the solar image moved across the entrance slit. The photographic plate was moved through a series of mechanical gears at a similar rate behind the second slit. The recorded slices of light then gradually built up to formed a complete photographic image of the solar surface and immediate surroundings.

Figure 4.5 shows an early (1892) SHG image taken in CaK.

Deslandres' concept was very similar but made use of a two mirror cœlostat (see Sect. 11.1) which sent the light of the Sun towards an objective lens placed in an opening in the wall of the building. His 25 cm objective focused an image of the Sun on the entrance slit of the spectroheliograph.

The displacement of the objective lens caused the solar image to pass in front of the entrance slit. In time with this motion, a photographic plate slid behind a slit placed at the spectral line. The result was an image of the Sun taken at the given wavelength. An early CaK wavelength image from Deslandres' instrument is shown in Fig. 4.6.

For the first time, astronomers could now study the solar chromosphere. The early images were taken in the prominent lines of CaK and Hβ due to the blue sensitivity of the available photographic plates. You can imagine the excitement and amazement when these first images were recorded! The recording of the first Hα images had to wait until 1908. The necessary Panchromatic, red sensitive film was only invented in 1902. In 1903 Hale further improved his original Kenwood instrument and built the Rumford spectroheliograph (SHG) for use on the Yerkes 40″ (1 m) refractor. The Rumford SHG used two Voigtlander camera lenses and two 60° prisms which he thought improved the efficiency, giving a brighter spectrum and also reduced scattered light.

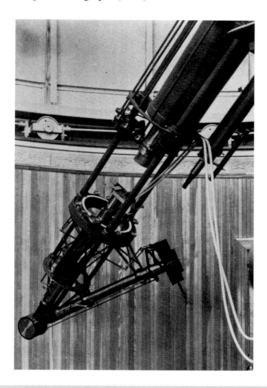

Fig. 4.4 Hale's original SHG as installed at his Kenwood observatory (Hale)

Many observatories around the world quickly picked up on this new method of observing the Sun. One of the first spectroheliographs to be constructed in Europe was set up by Paul Kempf (1856–1920) on the Grubb 7.5″ (190 mm) refractor at the Potsdam Astrophysical Observatory in 1904. In the same year the Solar Physics Laboratory at South Kensington, London constructed a spectroheliograph. This spectroheliograph incorporated a large siderostat (see Sect. 11.1) to feed sunlight to the instrument, which scanned the solar surface by moving sideways. The photographic plate was held separately and remained stationary as the image was built up.

Although quickly adopted by the professionals, there is no evidence that amateurs took up the challenge of imaging with the early spectroheliograph. The amateur was more attracted to the later visual adaptation—the spectrohelioscope (SHS).

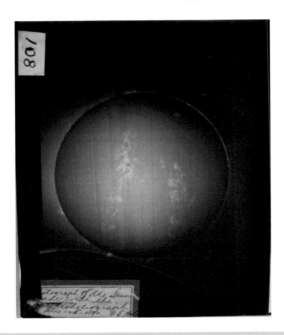

Fig. 4.5 Hale's 1892 CaK SHG image (Hale)

Fig. 4.6 Deslandres' Meudon image, 1897 solar disk in CaK (Meudon)

4.6 Persistence of Vision

The visual version of the spectroheliograph, the spectrohelioscope relies on the observer's persistence of vision to give the illusion of continuity and produce a complete image from a series of smaller image sections observed by the eye in quick succession.

The Thaumatrope, which was probably invented in 1825 by Dr. John Ayrton Paris (1785–1856), was the first of the early Victorian toys based on persistence of vision, and was the simplest in design. On one side of a small round board was drawn a bird; on the other was a cage (Fig. 4.7). When the board was held at the sides by two strings and spun, both images merged and the bird appeared to be in the cage.

The Zoetrope, or Wheel of Life, used the principle of arranging slits to view a series of images inside a rotating drum. Invented in 1834 by William George Horner (1786–1837), the images were drawn on a removable strip of paper, so the animations were changeable. The images were equally spaced around the inside of the drum, and the slits were spaced along with them, one slit per image. The viewer spun the drum and watched the animation through the slits (Fig. 4.8). This was perhaps the most popular and longest lasting of this type of toy.

A similar method of using moving slits to observe slices of the solar disk and perceive a whole image was the basis of the evolution of the spectroheliograph to the visual spectrohelioscope.

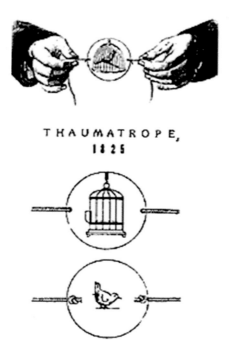

Fig. 4.7 The Thaumatrope (WIKI)

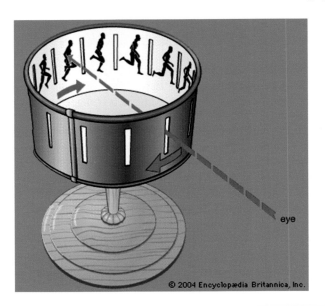

© 2004 Encyclopædia Britannica, Inc.

Fig. 4.8 The Zoetrope (Britannica)

4.7 Spectrohelioscopes (SHS)

By the mid 1920s Hale was investigating the possibility of adapting his photo-graphic spectroheliograph to visual observations, and in 1924 developed the SHS. One of the driving forces behind the idea of the SHS was described by Hale as follows:

> This method ... seems suddenly to bring to life the flocculi on the disk, which appeared fixed and inert on the photographic plate .... As the photographer using a spectroheliograph can-not see what is happening, he makes his exposures in a routine way and thus almost invari-ably fails to catch the successive phases of remarkable short-lived phenomena that often afford marvellous spectacles to the visual observer, who can pick out at a glance the most interesting and most active areas. Thus I have repeatedly seen with the spectrohelioscope the swift flow towards sunspots of masses of hydrogen larger than the Earth, adequately recorded only once with the spectroheliograph in a period of twenty years.

The original concept is shown in Fig. 4.9.

These original instruments targeted a solar diameter of approximately 2″ (50 mm)—this was the minimum size thought necessary to give adequate visual resolution. The telescope lens was a simple plano-convex lens of 18′ (approxi-mately 5.4 m) focal length. Two spherical mirrors, 13′ (4 m) focal length were used for collimating and imaging in what would be called a Czerny-Turner arrangement (see Sect. 6.3.4). A 15,000 l/in. (590 l/mm) speculum metal grating, approximately 2″ square (50 mm) was used to give the dispersion. There was no

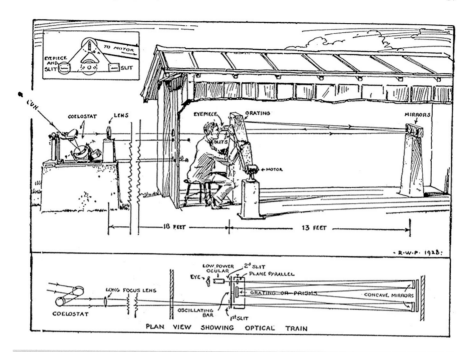

Fig. 4.9 Layout of the Hale spectrohelioscope (Ingalls, ATM Book1)

way such an instrument could be mounted on any of the usual equatorial mounts, so a system of mirrors was used to project the Sun's image onto the entrance slit (see Sect. 11.1). The overall size of the instrument was therefore very large and cumbersome.

Like the original spectroheliograph, the entrance and imaging slits of the spectrohelioscope (0.004″, 0.1 mm or 100 μm) were linked. In this case a wobble bar driven by a small electric motor held both slits approximately 3½″ (90 mm) apart and oscillated ±3/16″ (5 mm). This provided 30–40 oscillations per second, enough for the eye's persistence of vision to see a continuous image. Later, rotating square prisms known as Anderson prisms were mounted in front of a pair of fixed slits to provide the continuous image.

Ingalls' Amateur Telescope Making, Book 1 provides a detailed description and comprehensive construction details for both Hale's early spectroheliograph and his later spectrohelioscope. Although Hale may have been very enthusiastic about the spectrohelioscope and its capabilities, it doesn't appear to have been adopted by the professionals, whom we can only assume, were happy to continue with their photographic recording of the solar chromosphere with the well established spectroheliograph. As Fred Veio recounts in his book on the spectrohelioscope instrument, many amateurs however picked up on the Ingalls book article and proceeded to build the next generation of instruments.

Most of the amateur spectrohelioscope designs of the twentieth century were based on variations of the Hale design. These mechanical instruments are still in use today by amateurs and are discussed in more detail, with examples, later in Chap. 11. More recently, the advent of the webcam in the early 2000s allowed the amateur to simplify and refine the non visual spectroheliograph design into the successful digital instrument we see today. Interestingly, all amateur SHGs are still homemade DIY instruments. Notwithstanding Meade Instrument's patent raised in 2007 for a version of a digital SHG, no manufacturer has ventured into commercial production.

4.8   The Digital Spectroheliograph

Back in 1992 there was a watershed moment that became a definite game changer. It was the year of the Barcelona Olympics, Shuttle Endeavor's first flight, Hurricane Andrew, the founding of the European Union—and the first time a digital device was mated to a spectroheliograph!

On this momentous occasion Philippe Rousselle, working from his home in Metz, France, used a linear CCD obtained from a small hand scanning device to replace the traditional second slit in his prototype spectroheliograph and programmed an ATARI 1040 computer to interface and control the CCD. This allowed the recording of a narrow strip of the solar spectrum. These strips, when assembled sequentially with many others, built up an extremely narrowband image or spectroheliogram of the Sun. The digital revolution had begun.

From these early tentative steps, it was almost 10 years until webcams became readily available and could be applied to the recording of the solar spectrum. The advent of the webcam brought with it other problems however. Philippe had used a linear array CCD. This 2048×1 array of 14 μm pixels, when properly positioned, recorded the narrow strip of light 1 pixel wide in the target wavelength. To produce an image, these individual strips were then laid side by side, one after another, to form a continuous image of the Sun. It is a relatively simple imaging process.

The webcam by contrast uses a 2D array of 640×480 pixels. The horizontal size of the chip therefore records a continuous image of the solar spectrum. The width of the chip limits the extent of the recorded wavelengths, which depending on the optical arrangement of the spectroheliograph could extend over hundreds of Angstrom. The webcam or a similar fast frame camera can produce, as the entrance slit of the spectroheliograph scans across the solar image, a multi frame AVI video file. Depending on the scan rate the file can therefore contain over a thousand frames. It is anticipated that the target absorption line is recorded in a few pixel columns within each frame at exactly the same position along the CCD X axis. The pixel width of the line will depend on the capabilities of the spectrograph used.

A process was needed to separate the specific wavelength of interest, which effectively meant replacing the second slit used by Hale in the original SHG with a virtual software slit.

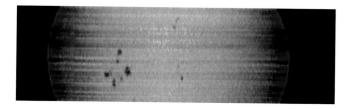

Fig. 4.10 Early digital SHG examples (Defourneau)

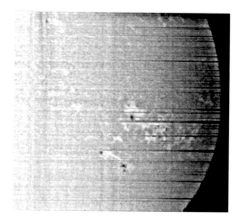

Fig. 4.11 Hα Image (Buil)

The software challenge was met by another Frenchman, Daniel Defourneau, working in the outer suburbs of Paris. In early 2002 he used one of the first commercial webcams to record the solar spectrum and also developed a software package to strip the required image columns from the webcam spectral AVI file. He then re-assembled the strips, laid side by side, to form a spectroheliogram mosaic of the solar image (see Fig. 4.10). The digital era had begun.

Later the same year Christian Buil obtained his first SHG images (Fig. 4.11) using a similar webcam on a classical spectrograph with a FS-128 refractor, stopped down to an aperture of 50 mm.

Many amateurs quickly picked up on the idea and successfully produced solar images from digital spectroheliographs. All of the solar features and magnetic solar effects mentioned elsewhere in this book can be successfully imaged with the digital spectroheliograph, making it an ideal instrument. Some recent spectroheliograms showing results are included in later chapters as we go on to discuss the process in more detail.

Further Reading

Hertschel, K.: Mapping the Spectrum, Oxford University Press (2002)
Lockyer, J.N.: Proceedings of the Royal Society, **15**, 256 (1866)
Meadows, A.J.: Science and Controversy, 2nd Ed., Macmillan (2008)
Maunder, E.W.: Sir William Huggins and Spectroscopic Astronomy. TC&EC Jack (1913)
Huggins, W.: The scientific Papers of Sir William Huggins. Wesley & Sons (1909)
Jackson, M.W.: Spectrum of Belief: Joseph Von Fraunhofer and the Craft of Precision Optics. MIT Press (2000)
"Étude des gaz et vapeurs du soleil," in L'astronomie (Dec. 1894)
"Recherches sur l'atmosphère solaire. photographie des couches gazeuses supérieures…," in Annales de l'Observatoire d'astronomie physique de Paris, **4** (1910).

Webpages

http://adsabs.harvard.edu/abs/1894BuAsI..11…55D
http://adsabs.harvard.edu/abs/1929ApJ….70..265H
https://archive.org/details/philtrans07911135
http://spectrohelioscope.org/
http://en.wikipedia.org/wiki/Zoetropehttp://www.astrosurf.com/buil/lhires2_Sun/first.htm
http://www.google.com/patents/US7209229
http://adsabs.harvard.edu/abs/1905ApJ….21…49K
http://adsabs.harvard.edu/full/1903PYerO…3….1H

Chapter 5

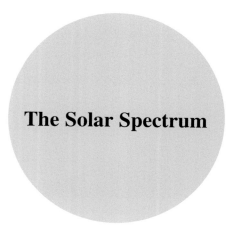

The Solar Spectrum

The importance of observing the solar spectrum was discussed previously, but it is worth highlighting how the new discoveries achieved using the spectroscope quickly enhanced our knowledge of the Sun. By the early 1900s the advent of the spectroheliograph became one of the most important solar research tools available. With the digital spectroheliograph now on the scene, the spectrograph component is newly providing the means of dispersing the incoming sunlight into the spectral band. The amount of detail which can be recorded within the spectrum is dependent on many parameters, but basically is controlled by the entrance slit gap and the type of diffraction grating used. Like telescopes in general, the resolution which can be achieved from the spectrograph is the key to success. Higher resolution allows more detail to be recorded in and around the target wavelength.

Before considering the science contributions which can be made by the amateur using the digital SHG, we need to re-visit, in general terms, the nature of the solar spectrum. It contains key features which can be used not only to process detailed solar images but which also allow us to observe and analyze the dynamics and magnetic effects critical to improving our understanding of the Sun.

5.1 Introduction

The solar spectrum has been observed and recorded from the time of Newton. Fraunhofer in the early 1800s not only noted the prominent absorption lines but also the fact that the light intensity distribution varied from the blue to the red parts of the spectrum. He correctly showed that the peak of the intensity is around the green region of the spectrum and that there was a longer extended tail reaching all

© Springer International Publishing Switzerland 2016
K.M. Harrison, *Imaging Sunlight Using a Digital Spectroheliograph*,
The Patrick Moore Practical Astronomy Series, DOI 10.1007/978-3-319-24874-5_5

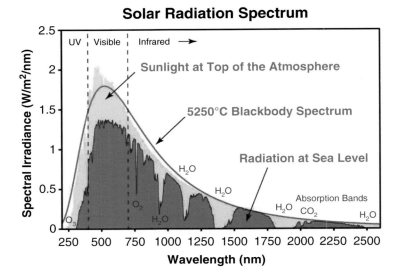

Fig. 5.1 solar energy distributions (WIKI)

the way beyond the red end of the visual spectrum. John Herschel later demonstrated using alcohol impregnated strips that there were still substantial heat energies being received in the solar spectrum well beyond the visual red region. He called this the infrared region. We can now accurately determine the distribution of the solar energy, which peaks around 5000 Å (500 nm), presents $< 10\%$ in the UV, $< 50\%$ in the visual $(4000 - 7000\text{Å})(400 - 700\text{nm})$ and $< 50\%$ in the infrared. At the Earth's surface the solar energy received is approximately $1000\,\text{W/m}^2$ (Fig. 5.1).

Given these numbers, the total energy received when observing the Sun with amateur telescopes can be dangerous and, as highlighted in Chap. 3, requires suitable filters to achieve safe limits when imaging.

The situation is slightly different when we consider a spectrograph mounted on a reasonably sized telescope. The only element which can receive the un-filtered solar energy is the entrance slit. The narrow slit gap employed, at only around 20 μm, limits the transmission to a very narrow strip of the solar image, and this energy is subsequently dispersed even further by the grating (by a factor of $> \times 10$ to the camera. Viewing or imaging at say, Hα wavelengths using a SHG is therefore very safe.

The smooth spectrum intensity curve is interrupted by the various absorption regions due to the Earth's atmosphere. These Telluric bands not only reduce the total energy received at the surface of the Earth, but provide for the imager, fixed reference lines in the spectrum. These can then be used to calibrate the plate scale Å/pixel and the absolute wavelengths recorded, allowing for comparisons between spectral results. The wavelengths of the important Telluric lines are listed in Appendix A.

5.1.1 Black Body, Planck and Wien Energy Distribution

Kirchhoff originally proposed that all bodies would radiate their energy in a similar manner, following a curve, the shape of which depended on their temperature. This was his black body theory. Planck and Wien later showed that the energy distribution could be mathematically modeled by the equation:

$$\lambda_{max=} 28,978,200 / T$$

Where λ is the peak emission wavelength (in Å) and T the temperature in degree Kelvin (°K).

The solar energy distribution recorded then allows us to easily calculate the surface temperature of the Sun. It closely follows the black body curve for a temperature of 5600 °K and is illustrated in Fig. 5.1.

5.2 Spectral Features in the Solar Spectrum

More than 90 % of our knowledge of the Sun has been acquired by analysis of the solar spectrum. The analysis of the distribution, the strengths and shapes (deep/shallow cores, extended wings etc.) of the various absorption lines visible in the solar spectrum has provided the astronomers with a detailed understanding of the composition and physical nature of the Sun. We can now confidently determine the chemical make-up, the pressures and temperatures, magnetic fields and monitor the material movements in and around the active areas and outer atmosphere of the Sun.

Using the digital SHG the amateur can replicate most of the important scientific discoveries which helped determine the dynamics of the solar features, from the surface photosphere to the higher altitudes of the chromosphere, as well measuring the varying strength of the solar magnetic field in and around active areas and record data which until recently was only accessible to the professionals.

When you look at the solar spectrum the absorption lines and bands appear to be everywhere. Some are much fainter than others and some are much wider than others. This variance in the size and visibility of the individual lines is influenced by the amount of the element in the Sun and the local solar conditions like pressure and temperature.

The Henry Draper stellar classification system (HD) was developed by Pickering (1846–1919) and his dedicated team from the analysis of the spectra from over 230,000 stars. This classification system gave rise to the well known spectral sequence OBAFGKM. The blue bright, high temperature stars are found at the beginning of the sequence as OB type stars, while the sequence ends with the type M cooler red giants.

The Sun is a typical G2V type star in the HD classification, as determined by the surface temperature of around 5800 °K and the dominant absorption features seen in its spectrum. The typical G2V star shows weak hydrogen lines, strongly ionized calcium lines and strong neutral metal lines. The CaH and K lines are the most prominent features visible in the solar spectrum. The hydrogen lines or the Balmer Series extending from 3600 Å through to 6563 Å are still easily visible, as are the numerous Iron (Fe), Magnesium (Mg) and Sodium (Na) lines. It was as a direct result of the limitations of the early photographic materials that the first investigations of the solar spectrum were carried out in the CaK and Hβ wavelengths. Later, as the photographic film sensitivities improved, this work was extended to the Hα wavelength. Most of the narrowband solar images obtained by the amateur, due to the limited availability of suitable filters, have been acquired in the CaK and Hα wavelengths. The digital SHG now allows the solar imager to record their images in almost any resolution and wavelength.

5.3   Formation of the Absorption Lines

The Sun is made up of 99 % hydrogen and < 1% helium, with all the other elements making up the balance.

The solar surface or photosphere, this is a thin layer of the solar atmosphere probably only 200 km deep where the majority of the solar energy is radiated/emitted. The light from the photosphere surface presents a continuous emission spectrum, the intensity of which follows a black body temperature curve for 5780 °K.

Above the photosphere lies the chromosphere. This region was previously only visible during rare total solar eclipses, but can now be regularly observed using narrowband filters. The temperature in the chromosphere varies from 6000 °K at the photosphere interface, falling to 4000 °K at a height of 500 km before steadily rising to 7000 °K at altitudes of 2000 km, then rapidly rising up to > 30,000 °K at the outer reaches, some 25,000 km above the photosphere (see Fig. 3.2).

As a result of the early total solar eclipse observations, astronomers developed the concept of the reversing layer in an effort to explain the bright red emission lines of the flash spectrum. This was thought to be a narrow cooler layer or region just above the photosphere in which emission lines were generated in accordance with the Kirchhoff laws. The brief flash spectrum is actually a result of the moon exposing only the upper and then the lower parts of the chromosphere without interference from the underlying photosphere. The intensity of the emission lines of the elements found in the absorption spectrum—hydrogen calcium, iron magnesium—is dependent on the amount of the chromosphere covered by the moon. At the outer reaches of the chromosphere the temperatures are high enough (20,000°) to excite the helium atoms, giving rise to the helium (D3) line which is not normally observed easily in the solar spectrum.

The dark absorption bands evident throughout the solar spectrum are actually caused by the differing emission and absorption of the various elements within the chromosphere and are affected by factors like temperature, abundance of the element, collisions between the atoms and photoelectric effects. As Zirin puts it,

"If we substitute the phrase "lower excitation temperature" for "cool" we will have the right picture".

The different parts of the absorption lines are formed at different heights due to the changing temperature and conditions within the chromosphere (see Fig. 3.2). Here we see that the calcium (K3) core absorption occurs around 6000° and an altitude of 1900 km and the greatly extended calcium wings (K1) are formed in the lower temperature regions around a height of 500 km. Similarly the wings of the Hα line are formed around the base of the chromosphere $(< 400\,\text{km})$ whereas the core region actual is formed at much higher regions, at about 1500 km.

By just concentrating on the CaK and Hα wavelengths the amateur can map the chromosphere from just above the photospheric surface all the way up to a height of 2000 km. The amount of image detail which can be recorded at each wavelength will be determined by the bandwidth resolution of the filter used. Obviously a narrow bandwidth will allow access to the deeper regions of the line core and provide higher contrast results. Typically a FWHM bandwidth of 0.7 Å is employed in the commercial Hα filters. If accurately centered on the Hα wavelength then this filter would allow imaging down to around 50 % the core depth (Fig. 3.15), adding a double stack (DS) filter would reduce the bandwidth to 0.5 Å and get deeper into the core thereby enhancing the recorded detail.

The imaging results at the core and out into the wings $(\pm 1\,\text{Å})$ will dramatically show the differences across almost 1000 km of the height of the chromosphere; the detail seen in the wings being at a much lower altitude. The visibility of the chromospheric features is illustrated in Fig. 3.16. Note that the highly magnetic cores of the sunspots can be recorded at all heights within the chromosphere.

The digital SHG offers the ideal tool for recording the chromospheric detail.

Not only can we select any central wavelength (CWL) associated with almost any element in the core or wing regions with a degree of accuracy unobtainable by most filters, we can also select the recording bandwidth from 0.2 Å upwards with equal accuracy. This precision as we will see in Sect. 5.5 allows the accurate determination of Doppler shift and magnetic activity.

The details of the associated atomic quantum theory and the mechanics of the processes which give rise to the various absorption lines and their relative strength are very mathematical and too complex to be included in this work. The interested reader is directed to Zirin's and Phillip's books where the subject is given extensive coverage.

5.4 Atlas of Solar lines

There are many atlases of the solar spectrum available to the amateur. These allow positive identification of the various lines recorded by the spectrograph. The spectral atlas, Fig. 5.2, from the University of Liege has been used widely by the spectroscopy community and meets most of the amateur's needs.

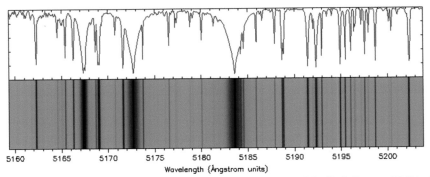

Data from "Photometric Atlas of the Solar Spectrum from 3000 to 10,000 A" by L. Delbouille, L. Neven, and C. Roland
Institut d'Astrophysique de l'Universite de Liege, Observatoire Royal de Belgique, Liege, Belgique, 1973
Image copyright © 2002 by Ray Sterner, Johns Hopkins University Applied Physics Laboratory

Fig. 5.2 Image of solar spectrum around the Mg lines (Liege Atlas)

The other popular solar spectral atlas is the BASS2000 atlas from Meudon (Fig. 5.3). Meudon also provides a downloadable PDF file (4.3 Mb) which covers the whole solar spectrum (from 3800 Å to 8700 Å). This highly detailed version includes annotations and Landé factors for all the main absorption lines. Both these atlases allow the user to zoom in on a target wavelength and are very user friendly.

We can use the known absorption lines detailed in the atlas to provide calibration wavelengths for our spectra and hence establish the dispersion (Å/pixel). This information, complemented by the use of reference lamps (discussed later), then gives us the basic information we need to use in selecting the spectral data for subsequent image processing.

5.5   Doppler and Other Spectral Effects

The rotation of the Sun and the dynamics of various features like filaments and prominences cause significant changes in observed velocities. These can be recorded by taking the Doppler effect into account. Also, the solar magnetic field can cause absorption lines to appear doubled or sometimes tripled in the Zeeman effect. Other interesting effects which can be observed are the Evershed effect where the inflow and movement of material in sunspots can be recorded and the Severny moustaches associated with flares.

HIGH RESOLUTION SOLAR SPECTRUM

UV: from 670 Å to 1609 Å (SOHO/Sumer), resolution 0.04 Å
Visible: from 3000 Å to 10000 Å (Jungfraujoch), resolution 0.002 Å or 500 pixels/Å
Infra Red: from 10000 Å to 54000 Å (Kitt Peak), resolution 0.004 cm-1 (varies from 0.004 Å at 10000 Å to 0.1 Å at 50000 Å)

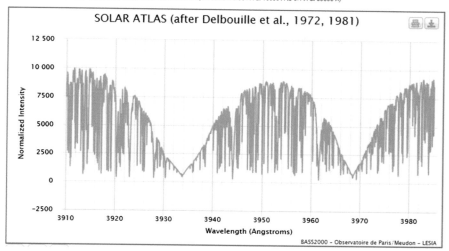

Fig. 5.3 CaK and CaH region (BASS2000)

5.5.1   Doppler Effect

When using a high dispersion spectrograph (see Sect. 6.5 for more on this) we are able to record and measure the Doppler effect. These results then allow the amateur to confirm the movements of material and features like filaments and prominences on the Sun.

As material moves towards us, the wavelength of any spectral feature will be displaced towards the blue side of the spectrum; likewise material moving away from us will show a spectral shift towards the red. Although the CWL will shift, the general shape of the individual absorption lines will remain unchanged. This radial velocity is usually termed line of sight velocity when applied to the Sun.

The formula for Doppler shift is:

$$\Delta\lambda \,/\, \lambda = v \,/\, c$$

Where $\Delta\lambda$ is the spectral shift (in Å), λ the stationary wavelength (in Å), v the velocity of recession or approach—a negative value (in km/s), and c the velocity of light $\left(2.998 \times 10^5 \, \text{km} \,/\, \text{s}\right)$.

Considering a resolution of $0.2\,\text{Å}$ at Hα, this would record a velocity of:

$$v = \Delta\lambda / \lambda \times c$$

Line of sight velocity, $v = 0.2 / 6563 \times 2.998 \times 10^5\,\text{km / s} = 9.1\text{km / s}$

Some of the chromospheric features like flares, filaments and prominences can exhibit large line of sight velocities, up to 100 km/s in some cases and a Doppler shift of the Hα line by $> 2\,\text{Å}$. To observe these features requires the filters used need to be tuned in CWL to accommodate this Doppler shift. Zenter illustrates this well in Chap. 10, Fig. 10.4 where the Doppler shift of the CaH line in a prominence is recorded.

5.5.2   Zeeman Effect

The Zeeman effect is caused by the splitting of absorption lines in high magnetic fields. The splitting of the lines can be observed with high resolution spectrographs and the results used to confirm and map the magnetic activity in and around sunspots.

Our Sun is a highly magnetic star. The magnetic field flux can vary from 0.001 T to over 0.4 T in some sunspots. The magnetic field also varies from the equator, were it is strongest, to the poles where it's only 1/6 the equatorial strength.

1 Tesla is equivalent to: 10,000 (or 10^4) G (gauss), used in the old CGS system.
    Thus, $10\,\text{kG} = 1\text{T}$ (Tesla), and $1\text{G} = 10 - 4\text{T}$.)
5 mT (0.005 T)—the strength of a typical refrigerator magnet
25–65 mT—the strength of the Earth's magnetic field
1–2.4 T—coil gap of a typical loudspeaker magnet

The Sun's major component of magnetic field reverses direction every 11 years. That makes the magnetic period about 22 years, resulting in a diminished magnitude of magnetic field near reversal time. During this dormancy, the sunspot activity is at maximum. Many amateurs may have read briefly about the Zeeman effect. Basically, a strong magnetic field can cause a separation of the observed absorption line into two or three individual lines.

For users of the digital SHG, recording the Zeeman effect can provide a challenging and satisfying project.

The Zeeman effect (Fig. 5.4) occurs when the energy levels of an atom are split by a magnetic field. This was first discovered by Zeeman (1865–1943) in 1896, who was awarded the Nobel Prize for Physics in 1902 for his work. One of the early results of Hale's SHG was obtaining confirmation that the Zeeman effect was caused by the solar magnetic field.

The Zeeman effect is controlled by the quantum number Mj of the atom being observed, the quantum number being the projection of the total angular momentum J on the direction of the magnetic field. The transition or splitting of the absorption line is determined by the strength and direction of the magnetic field and the atomic structure of the atom being considered.

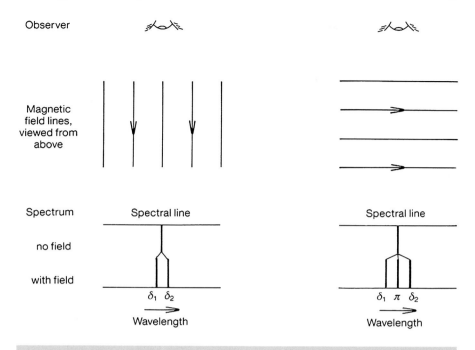

Fig. 5.4 Zeeman Pattern (Phillips)

The transitions are also sensitive to polarization of the light. Linear polarization, π polarization, drives the $Mj = 0$ transitions, whereas circular polarization, σ polarization, gives the $Mj = \pm 1$ transitions.

The, σ transitions are shifted in wavelength by:

$$\Delta\lambda_H\left(\text{Å}\right) = 4.7 \times 10^{-9}\, g\lambda^2 H$$

where H = magnetic field strength in Tesla
g is the Landé g-factor

The Landé factor can vary from 1 to 3 depending on the atomic structure.

Example: Based on the 5250 Å line $\left(g = 3\right)$, the splitting in a 0.1 T field is only 0.032 Å. This would require a spectroscope working at $R = 164,000\left(\lambda / \Delta\lambda\right)$ to fully resolve!

When working with magnetic fields at an atomic level, the professionals work in wave numbers, $\tilde{v},(\tilde{v} = 1 / \lambda \, \text{cm}^{-1})$, the number of wavelengths per unit distance equivalent to the number of cycles per wavelength, where λ is the wavelength.

$$\tilde{v}\,\text{cm}^{-1} = 10,000,000 / \lambda\,\text{nm}$$

This is sometimes termed the spectroscopic wave number. (I.e. Hα at $6563\text{Å} = 15236.93\text{cm}^{-1}$)

The units of the Bohr magnetron $(\mu B) \times \mu B = 0.467\text{cm}^{-1}\text{T}^{-1}$ are also used.

With a polarizer oriented to pass light polarized along the field direction, the frequencies expected to be observed are νo, and $\nu o \pm g \mu BT$. Where νo is the frequency of the line in zero field, g is the Landé g-factors of the two levels involved, T the magnetic field strength in Tesla.

Example:
Ca I, wavelength number $= 23652.30\text{cm}^{-1}, g = 1$, and a magnetic field of 1 T. The splits would be $= \pm 0.467\,\text{cm}^{-1}$ Converting this back to a wavelength, $\nu o = 4227.9\text{Å}$

And the split $= 0.1\text{Å}$, needing a resolution of at least $R = 43,000$

Amateurs can record the Zeeman effect where the field strength is above 0.2 T, generally in sunspots. Fine structured Fe lines make good targets but need a very high dispersion/resolution. Chapter 10 has further details, and a list of suitable Zeeman target lines is given in Appendix A.

5.5.3   *Magnetography and Magnetograms*

One of the earliest and significant solar discoveries of the twentieth century was the powerful effects of the solar magnetic fields. This was first investigated by Hale in 1908 using the new solar tower telescopes installed at Mt. Wilson. He subsequently confirmed the Zeeman effect and determined that pairs of sunspots leading/western vs. following/eastern had opposite magnetic polarity. He also established that the bipolar nature of the sunspots varied in each solar hemisphere, with the northern hemisphere being the reverse of the southern. By 1913 Hale also found that the sunspots of the new solar cycle which had just commenced had reversed polarity from the previous cycle and thereby confirmed the 22 year magnetic solar cycle.

The bipolar magnetic nature of sunspots can be recorded by the amateur by obtaining magnetograms, maps of the magnetic field with a modified digital SHG. Preparing magnetograms requires additional optics elements to be added into the SHG. Basically, a quarter wave retarding plate which can be rotated through 90° and a linear polarising filter need to be incorporated into the optical path between the telescope and the entrance slit. Figure 5.5 shows the arrangement.

These added optical components allow the separate Zeeman lines (δ_1 and δ_2 in Fig. 5.4) to be selected and imaged according to the angular position of the quarter wave plate. Magnetic fields of 0.01 T (100 G) and above have been successfully recorded using this method.

By taking two sets of SHG scans, one centered on the Zeeman blue (δ_1) and the other on the red line (δ_2) with different quarter wave plate positions (Fig. 5.5a, b), subtracting them (A1–A2 and B1–B2), then subtracting the Scan 2 result from Scan 1, a final composite image is generated. Figure 5.6 shows the centerline extraction position as a dotted line.

a 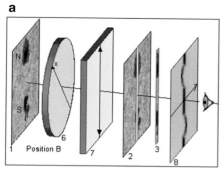 **b**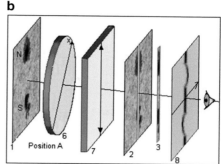

Fig. 5.5 (**a**) ¼ wave plate position A. (**b**) ¼ Wave plate position B. (*1*) Sun surface seen by the observer. Two sunspots of opposite polarity, N and S. (*2*) Image of the sunspots on the entry slit of the spectrograph. (*3*) Transmitted slit image into the spectrograph. (*6*) Quarter wave plate. Position A is used to record the blue (δ_1) Zeeman line image and Position B for the red (δ_2) (for an S polarity sunspot). (*7*) Linear polarizer at 45° to the axes of the quarter wave plate. (*8*) Resulting spectral image as recorded by the camera (after Rondi)

When these are stacked with 7 or more images used to improve contrast, the magnetic fields and direction are recorded. Then north magnetic direction will show as a brighter red (δ_2) line image and south the darker blue (δ_1) line image. Chapter 10 shows a magnetogram obtained using this method.

The line of iron FeI at 6302.5 Å is often used due to the high Landé factor $g = 2.5$. Professionals have "standardised" on line FeI 5250.2 Å $\left(g = 3.00 \right)$ for their magnetography. See Appendix A for a listing of suitable lines and their Landé factors.

5.5.4 Evershed Effect

In larger sunspots the movement of material in and around the sunspot can be recorded. The Evershed effect is seen when the entrance slit is placed over a sunspot umbra and penumbra: the absorption lines are seen to distort (Fig. 5.7) due to the Doppler shift and indicate a positive radial movement of material outward from the sunspot.

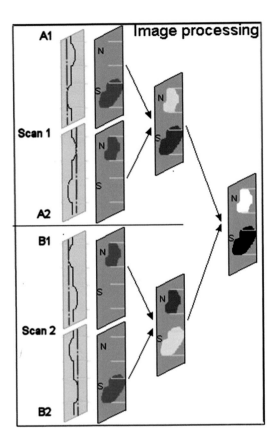

Fig. 5.6 Processing of the magnetogram images

5.5.5 *Severny Moustache (Flares and Ellerman Bombs)*

When the solar magnetic field constrained in a magnetic tube or loop collapses, it gives rise to the release of tremendous amounts of energy causing a flare to be recorded. When these bright points or flares are seen in Hα they are called Ellerman bombs and are usually observed in the strong magnetic fields surrounding sunspots.

When observed with the spectroheliograph, the emission spectrum of these flares shows the distinctive profile shape as illustrated in Fig. 5.8. The dotted line shows the general shape of the background continuum. For obvious reasons Severny called them moustaches.

The emissions extend in the blue and red wing of Hα by up to 5 Å, low in the chromosphere. The contrast against the underlying Hα absorption profile usually limits the recording of the emission in the wings to ±2.4 Å as indicated by the slit positions 1 and 3 in Fig. 5.8.

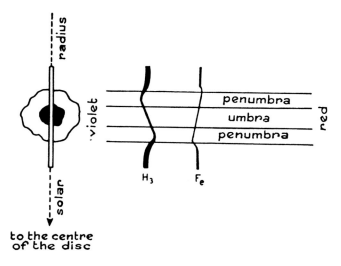

Fig. 5.7 Evershed effect. Slit placed radial across a sunspot (Abetti)

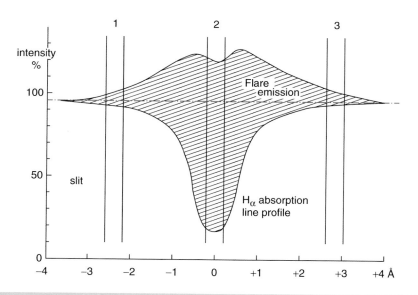

Fig. 5.8 Hα flare emission showing the moustache profile (Svestka)

Further Reading

Roth, G.D. (Ed): Handbook of Practical Astronomy, Springer (2010)
Robinson, K.: Spectroscopy – The key to the stars, Springer (2007)
Tennyson, J.: Astronomical Spectroscopy, World Scientific (2011)
Wilkinson, J.: New Eyes on the Sun, Springer (2012)
Harrison, G.R., Lord, R.C., Loofbourow, J.R.: Practical Spectroscopy, Prentice Hall (1948)

Webpages

http://www.google.com/patents/US7209229
http://adsabs.harvard.edu/full/1954ApJ...120...83G
http://adsabs.harvard.edu/abs/1959ApJ...130..366 L
https://www.newport.com/Optical-Filter-Construction/372468/1033/content.aspx
https://web.njit.edu/~cao/Phys780_Lecture07.pdf
http://articles.adsabs.harvard.edu//full/1966SSRv....5..388S/0000389.000.html

Chapter 6

Digital Spectroheliograph Design Basics

Overview

The digital spectroheliograph is nothing more than a combination of a suitable telescope, spectrograph and recording camera. It is an imaging system, not suitable for visual observing. The visual equivalent is the Spectrohelioscope, discussed more in Chap. 11.

When a SHG instrument is pointed at the Sun, any stable mounting can be used, and the solar image allowed to trail across the field of view. This image is focused on to the entrance slit of the spectrograph, which effectively scans the Sun's disk. The spectrum being produced is then recorded as an AVI video file which is subsequently processed to produce an image of the Sun (known as a spectroheliogram) in a selected wavelength.

This obviously has the benefit over commercial narrowband filters of allowing the preparation of multiple images at various wavelengths. All this is achieved from one video recording; an example would be the recording of the red/core/blue wing(s) of Hα.

Note: Solar observing can be dangerous. Not only to yourself but to the telescope and equipment you use. Use caution at all times.

© Springer International Publishing Switzerland 2016
K.M. Harrison, *Imaging Sunlight Using a Digital Spectroheliograph*,
The Patrick Moore Practical Astronomy Series, DOI 10.1007/978-3-319-24874-5_6

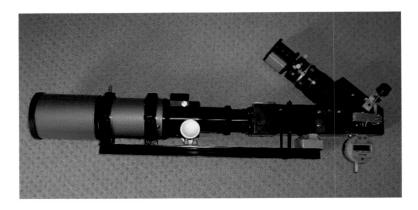

Fig. 6.1 Typical compact digital SHG

6.1 Design Basics: Key Components

The digital SHG is assembled from six components:

1. An imaging telescope, which produces an image of the Sun onto a slit.
2. An entrance slit which is a very narrow slit gap (usually 20 μm or less) and of sufficient length to capture the solar diameter (if possible).
3. A collimating lens, positioned behind the entrance slit to provide a parallel, collimated beam to the grating.
4. A reflective grating which disperses the light into the visible spectrum.
5. An imaging lens, to refocus the spectrum.
6. A recording camera usually a mono fast frame video/webcam camera, which produces an AVI file.

These components when combined together allow the recording of the spectrum of solar disk over a span of wavelengths. Each component is discussed in detail in the following sections. Using suitable software the wavelength of interest (say Hα) can be extracted from the AVI file, and then re-combined and presented as a mosaic covering the scanned region of the solar disk, forming a complete image.

Figure 6.1 shows a typical basic digital SHG design. This example is based on a Skywatcher ST80, 80 mm aperture, 500 mm focal length telescope, a classical spectrograph fitted with a 1200 l/mm grating and a DMK51 imaging camera.

The description and construction details of two small SHG's based on these designs notes are given in Chap. 7. The software currently available to the amateur is presented (with suitable user notes and illustrated with comparison spectroheliogram results) in Chap. 8. In Chap. 9 we will present construction details of some of the digital SHG designs being successfully used by amateurs around the world which the novice will find very informative and useful. For now, the following section reviews the design parameters of a suitable telescope for use in the digital SHG.

6.2 The Telescope

Many different solutions have been adopted by the amateur in telescope design, from standard photographic lenses to sophisticated reflective mirror systems. We will briefly discuss the common issues to be considered and addressed in order to reach a successful outcome.

The final choice of telescope/spectrograph design will be guided by the user requirements for resolution, size and complexity of the overall instrument, but before considering the specific optics used in a digital SHG, we need to familiarize ourselves with the basics of diffraction and resolving power.

6.2.1 Airy Disk, FWHM and PSF

All stars but the Sun are so far away that they appear as mere pinpoints of light to the naked eye. When we observe a star through a telescope at high magnification, it is immediately obvious that the star image appears as a small disk of light surrounded by faint rings, rather than a discrete point. So what turns this infinitely small pinpoint of light into a shimmering disk?

Airy in the late 1820s investigated the effects of diffraction and found that a perfect star image would form a disk (now known as an airy disk) that is brighter in the middle and surrounded by a series of concentric rings. Some 80 % of the light is concentrated in the disk (Fig. 6.2), the rest being spread out into the various rings.

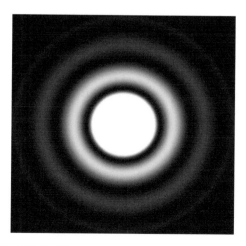

Fig. 6.2 Airy disk

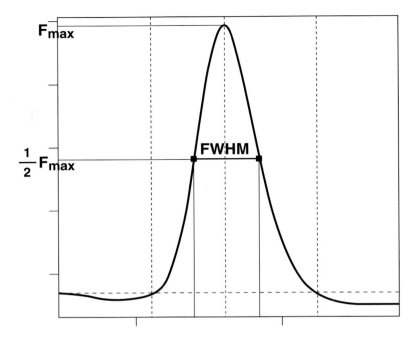

Fig. 6.3 FWHM of an intensity curve

The size of the disk is defined by the position of the first dark minimum between the disk and the first ring. The radius of this disk is defined as: $Sin\theta = 1.22\lambda/D$

$$\text{For small angles } Sin\theta \approx \theta \, (radians)$$

where λ is the wavelength of the light and D the diameter of the objective.

Therefore the angular size of this disk will vary with telescope aperture. The larger the telescope aperture, the smaller the disk size. The linear size of the disk in microns using a telescope with a focal length F is:

$$\text{Airy disk diameter} = 2.44\lambda F \, / \, D \, (\mu m)$$

A 200 mm f8 telescope would give a diffraction disk of 11 μm, whereas with a 400 mm f4 system it would be 5.5 μm.

As the intensity distribution in the Airy Disk approximates a Gaussian curve, 40 % of the light is contained within 50 % of its diameter. The term Full Width Half Maximum (FWHM) is widely used as a measure of the image size based on this Gaussian intensity distribution as illustrated in Fig. 6.3.

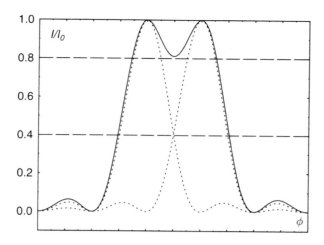

Fig. 6.4 Rayleigh limit

Typically for telescopes the Rayleigh limit is used to evaluate the optical performance. When two Airy disks are seen separated with the peak of one sitting in the first dark ring of the other they are said to be at the Rayleigh limit, a basic measure of the resolving power of the telescope or lens. At this separation, the FWHM disks are just touching each other (Fig. 6.4). We can therefore use the FWHM measure as one of resolution.

The Airy disk/Rayleigh limit gives:

$$\text{Resolution}\,(\text{arc second}) = 125\,/\,\text{Aperture}\,(\text{mm})$$

This infers a resolution, for a 200 mm aperture, of 0.63 arc second.

The real diffraction images produced by an optical system are defined in terms of the Point Spread Function (PSF). This basically provides a mathematical conversion equation between the ideal and real world outcomes. The Airy disk represents the result of the PSF for a single perfect objective. Multiple lenses or mirrors in the optical system each contribute their own PSF to the final resolution which can be obtained by the instrument. We usually use the Airy disk as a measure of the best image we can expect to produce. Unfortunately very rarely is this achieved due to other factors like the seeing conditions.

The same thing happens in a spectroscope. An emission spectral line, say from a neon reference lamp, has a minute bandwidth, maybe <0.01 Å wide; effectively a very narrow intense line (Fig. 6.5, A). When recorded by a spectroscope even with perfect optics and no aberrations it will still appear as a broad line with an intensity curve similar in profile to the Airy disk (Fig. 6.5, B).

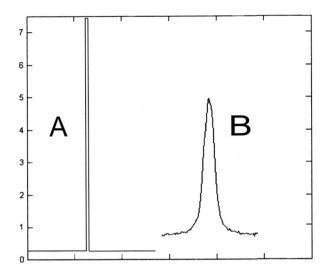

Fig. 6.5 Line intensity (*A*) vs. actual profile (*B*)

The overall PSF of the optical system will therefore define the resolution, both for the imaging telescope and the spectrograph. In reality the spectrograph resolution is measured in terms of the FWHM obtained from a reference lamp line profile (see Sect. 7.4 for more).

The neon lamp spectral profile shown in Fig. 6.6 demonstrates the difference between the grating dispersion and the actual resolution obtained. The dispersion shown is 1.09 Å/pixel whereas the FWHM resolution, as measured, is 4.64 Å.

For a spectrograph the resolution is quoted as an R value, where R = λ/FWHM. This example gives an R = 1264 (5854 Å/4.64 Å).

Generally, the local seeing conditions will determine the spatial resolution obtained.

6.2.2 Telescope Design and Energy Rejection Filters

The telescope used can be either a refractor or modified reflector. In the case of a refractor a simple achromatic design is probably the safest to use. Apo triplets with oil-filled objectives have been known to fail due to heat build up in the objective/mounting cell. Likewise, refractors which incorporate secondary lenses near the final focus such as Petzval designs could suffer damage with the concentrated energy. Al Nagler, founder of Televue, has gone on record to say that the Petzval design of the Televue telescopes is safe to use for solar observing without additional safety filters.

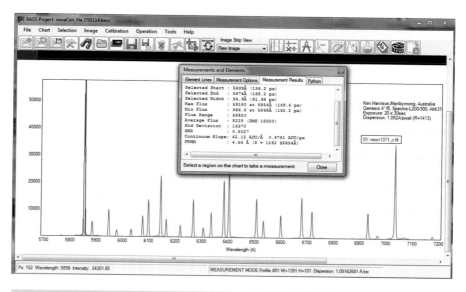

Fig. 6.6 Neon ref line showing measured FWHM (BASS Project)

For larger refractor apertures (>100 mm) adding an Energy Rejection Filter (ERF) suitable for the wavelengths you wish to record provides added safety. A good UV/IR rejection filter is a safe first choice. A Baader D-ERF for Hα work is an excellent solution. For other common wavelengths, the BelOptik Tri-ERF (CaK, Continuum, and Hα) is recommended.

Reflectors (Newtonian type), using uncoated primary mirrors, are a common choice. An SCT or any other compound reflector is not recommend unless good ERF filters are fitted securely to the front aperture as there is the distinct possibility of damage to secondary mirrors and their supports.

For the digital SHG the focal length of the telescope is an important factor. This will determine the size of the solar image produced at the entrance slit. Suitable Barlow lenses/Powermates can be incorporated into the telescope optical train to give a longer effective focal length.

Measuring the Focal Length of a Barlow Lens

You can quickly and easily measure the negative focal length of a Barlow lens by measuring the clear lens aperture and drawing a circle on a white card double the diameter. e.g. if the Barlow lens measures 28 mm, make the circle 56 mm. Project an image of the Sun which will appear as a large bright disk onto the card and move it back and forth until the image just fills the circle. Measure the distance to the Barlow lens. This measurement equals the negative focal length of the lens.

Most ×2 Barlows seem to be around −120 mm focal length; this is the distance the lens should be placed inside the prime focus of the telescope to give a parallel collimated beam output.

6.2.3   Solar Diameter vs. Focal Length

In its annual journey around the Sun, the distance from the Sun to the Earth varies. This distance is at a minimum in January and at a maximum in July. As a result, the apparent diameter of the solar disk can vary from 32.58 arc minutes (1955 arc seconds) to 31.51 arc minutes (1891 arc seconds). The average is 32 arc minutes, or 1920 arc seconds.

The projected size of the solar disk (in mm) is

$$\text{Diam.}(\text{mm}) = \text{focal length}(\text{mm}) \times 0.00948 (\text{Jan})$$
$$\text{Diam.}(\text{mm}) = \text{focal length}(\text{mm}) \times 0.00917 (\text{July})$$
$$\text{Diam.}(\text{mm}) = \text{focal length}(\text{mm}) \times 0.00934 (\text{Ave.}, 32 \text{ arc min})$$

Based on the average solar diameter, a 1000 mm focal length telescope would produce a solar disk almost 9.4 mm diameter. The plate scale would be $(32 \times 60)/9.34 = 205.5$ arc seconds/mm.

6.2.4   Surface (Spatial) Resolution vs. Seeing Conditions

Seeing conditions can vary with the time of day, usually better before noon and before the surroundings have heated up. and the local surroundings: large amounts of concrete in the immediate area can re-radiate the solar energy and cause air currents and atmospheric turbulence. The variations in ground temperature and the effects of the hot air rising can be minimized by observing from a tower or high platform. This is the theory behind the design of the Tower solar telescopes.

It would be a really good day for the amateur if seeing of 1 arc second could be achieved—it's usually more like >2 arc seconds.

Assuming best conditions, at 1 arc second seeing, then effectively we could record (1920)/1 = 1920 lines across the average solar diameter.

An aside: The solar diameter is 1,392,000 km, a 1 arc second image resolution would then represent 736 km on the solar surface.

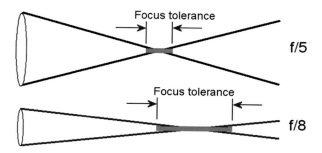

Focus tolerance

f/5

Focus tolerance

f/8

Fig. 6.7 Focus tolerances vs. focal ratio

6.2.5   Focus Tolerance

To get the best results from your instrument, a robust focusing mechanism must be included in the design for both the telescope and the spectrograph elements. It is a critical factor in achieving the best resolution. If we work to the Raleigh limit criteria of ¼ wave accuracy we find the allowable out of focus is approximately:

$$\Delta f = 4\lambda fr^2$$

Where fr is the focal ratio of the objective/camera lens, and λ the wavelength of the light being considered.

Straight away you can see that the focus tolerance in blue light will be half that of red light (3500 Å/7000 Å). To achieve the resolutions discussed previously (Sect. 6.2.1) we must be able to focus precisely the solar image (onto the entrance slit) and also the final spectral image on the recording camera.

The depth of focus, the allowable focusing error, depends only on the focal ratio of the telescope/lens system being used as shown in Fig. 6.7.

The critical focus zone" (CFZ) is sometimes quoted as:

$$CFZ = 2.2 \times fr^2 \ (\mu m)$$

This is based on perfect, collimated optics. A more realistic tolerance, based on achieving ¼ wave accuracy, can be expressed as:

$$CFZ = 4 \times \lambda \times fr^2$$

The data given in Table. 6.1 is based on this formula.

The final assembly and tuning of the digital SHG needs to be done accurately and the outcome must be capable of achieving this tolerance to perform with optimum success.

Table 6.1 Focus tolerances vs. focal ratio

Focal ratio	Focus tolerance (µm)		
	CaK	White light	Hα
15	354	495	590
10	157	220	262
8	101	141	168
5	39	55	66

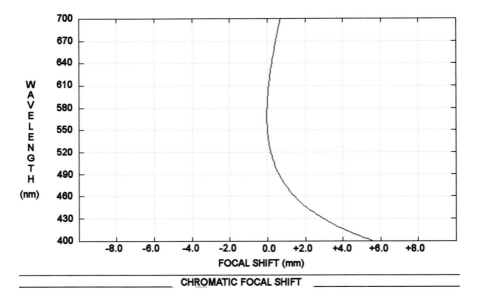

Fig. 6.8 Typical chromatic focal shifts—90 mm/850 mm focal length achromatic objective (Edmund Optics)

Reminder: The focus requirement is therefore twice as critical when imaging in the UV as in the red (i.e. CaK vs. Hα)

Chromatic Focal Shift: Refractors

All refractors suffer from chromatic aberration. The precise focal position will vary with wavelength. Figure 6.8 shows the chromatic focal shift with wavelength in a typical f9.5 refractor achromatic objective. This graph shows that the position of the focal point between 700 and 510 nm (7000 and 5100 Å) only varies by a maximum of 0.2 mm (230 µm). The CFZ (see Sect. 6.2.5) is approximately 220 µm (at 5100 Å) and 260 μm at Hα (6563 Å). This objective could therefore be used for white light and Hα imaging with little or no refocusing.

The situation changes dramatically when we consider the CaK wavelength (3934 Å, 393 nm). The graph clearly shows the focus point moving quickly away from the objective giving a longer focal length, and is approximately 5.7 mm from the Hα focus.

This focus change must be accommodated in the mechanical arrangement of both the achromatic telescope and any associated achromatic spectrograph optics being used. With reflective mirror optics there is no chromatic focal shift, all visible wavelengths being brought to a common focus.

Using Pre-filters as an Aid to Focusing

For larger apertures, as mention in Sect. 2.3, the BelOptik Tri-ERF is a good solution as an ERF. This filter by design also acts as a narrow band filter for both Hα and CaK wavelengths. Focusing these wavelengths using the BelOptik filter is therefore made much easier.

Two inch (48 mm filter thread) quality filters could also be used as "pre-filters", the Wratten 80b (Blue) for CaK and a Wratten 25A (Red) for Hα are popular choices. These filters should be positioned far enough in front of the focus to minimize heat build up, and vignetting.

Consider a 100 mm f10 optical system. The solar disk would be approximately 10 mm diameter at the focus, and the f10 light cone would allow a 2″ (48 mm clear) filter to be placed at a distance inside focus of 380 mm ((48−10) = 38 mm × 10 = 380 mm). These filters will reduce the light from the solar continuum and allow more precise focusing on the target wavelengths.

6.2.6   Mounting

The current design of the compact digital SHGs are usually lightweight and can be used on either Alt-Az or Equatorial mounts, with or without drives. The mount should be capable of securely holding the instrument with minimal vibration and allow some controlled motion to center the solar image and verify the scanning position on the entrance slit gap. Without drives it will take 1920/15 = 128 s for the solar disk to drift across the entrance slit gap

With an Equatorial mount, the RA drive can be adjusted to ×2 guide rate or faster, giving a scan time of 64 s. This will be discussed further when considering imaging cameras and frame rates. Chapter 9 details the construction and mounts used on various amateur instruments and shows the wide variety of solutions.

Larger SHGs which are more suited to static horizontal or vertical mounting can also make use of the original cœlostat mirror system to direct the solar image into the telescope (see Sect. 11.1 for more).

6.2.7 Summary: Telescope

- Apertures around 100–120 mm are recommended
- The primary image of the Sun will be approximately 1/100 the focal length
- Pre filters can be used to assist focussing for chromatic focal shift in achromatic refractors
- Seeing is more important than optical resolution with larger telescopes
- Scan rates can be changed when using an EQ drive system.

6.3 The Spectrograph Design Options

The type of spectrograph arrangement (entrance slit/collimator/grating/imaging system) used for solar work can be similar to those used by amateurs in stellar spectroscopy. The Shelyak LhiresIII and the JTW Spectra-L200 Littrow designs have been used in digital SHG construction. Having plenty of light to work with, compared to conventional night time astronomical spectroscopy, using far narrower entrance slit gaps than those normally found in standard astronomical spectroscopes, can be considered to our advantage. To achieve the best minimum bandwidth recorded in the final image we will find that large reflective gratings with l/mm >1200 combined with longer focal length optics are preferred. Suppressing excess light and reflections is mandatory for best contrast. The spectrograph section of the digital SHG is therefore usually different to the conventional astronomical spectrograph, being redesigned to better meet these criteria.

There are many spectrograph designs available to the amateur. All have their strengths and weaknesses.

6.3.1 Classical Design

The classical spectrograph is probably the easiest design for the amateur to construct (see Fig. 6.9). The collimating lens can be an achromatic doublet. The focal ratio of the collimator must match the focal ratio of the telescope and the diameter match the size of the grating being used. Spacing between the grating and the imaging lens should be kept as short as possible to reduce the likelihood of vignetting. The separation of the collimator lens from the imaging lens gives some design flexibility. If different focal length lenses are used, some de-magnifying giving a smaller solar image can be achieved, thus giving a better match to the CCD chip dimensions.

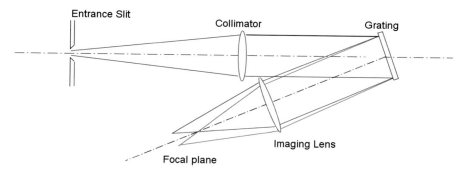

Fig. 6.9 Classical layout

6.3.2   Littrow

Using only one achromatic lens or spherical mirror for both the collimator and imaging lens in the Littrow design simplifies construction, but doesn't allow any control over the magnification factor between the slit image and the camera image. The total angle between the slit and CCD optical axes must be kept to a minimum to reduce astigmatism. On longer focal length designs, the pick-off mirror can be eliminated and the camera placed immediately above the entrance slit (Fig. 6.10).

6.3.3   Ebert–Fastie

The Ebert–Fastie design as shown in Fig. 6.11 is popular with SHG constructors. A large spherical mirror (M) is used for both collimating and imaging. There is some residual coma. In the usual in-plane configuration, the spectral lines produced are curved "smile", but this can be minimized by using an out of plane or up-and-over arrangement. It this case the spectral lines have a slant, which varies with wavelength. The minimum size of the mirror is approximately three times the width of the grating. The f ratio of the main mirror must be at least ×2 faster than the telescope f ratio i.e. an f10 telescope would need an f5 mirror.

6.3.4   Czerny–Turner

To reduce the optical aberrations, the Czerny–Turner design uses two separate spherical mirrors. This extra degree of freedom allows the tilt angles of the mirrors (angles α and β in Fig. 6.12) and their relative positions to be changed, thereby

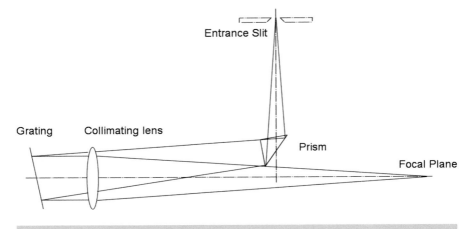

Fig. 6.10 Littrow optical layout

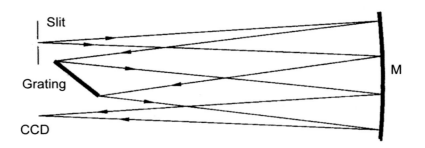

Fig. 6.11 Ebert-Fastie optical layout

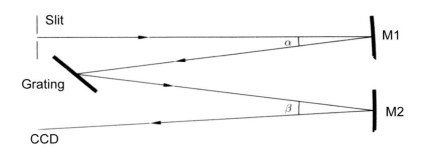

Fig. 6.12 Czerny-Turner optical layout

reducing astigmatism and coma. If the camera mirror has a shorter focal length than the collimator the resulting coma will be negligible. The collimating mirror (M1) should be slightly larger than the grating size being used and the imaging mirror (M2) large enough to collect the dispersed beam from the grating. At least the collimating mirror f ratio must match the telescope being used. The SimSpec SHG spreadsheet will help confirm the sizes required (see Appendix B).

6.3.5   Summary: Spectrograph Optical Arrangements

- Longer focal length systems give higher dispersion
- Use large high l/mm blazed gratings to improve resolution
- Gratings can be used in second order to increase dispersion

6.4   The Entrance Slit

The slit design and its length need to accommodate the size of the solar disk and the eventual size of the spectral image produced by the imaging system. Although the solar image from the telescope may be, say 9 mm diameter at the entrance slit, by changing the focal lengths of the collimator/imaging camera lens it can be recorded at smaller dimensions to better suit the CCD i.e. reduced down to 4 mm (see Sect. 6.7)

The energy being transmitted by the telescope to the solar image focused on the slit can cause some heating and possible distortion of the slit. A 100 mm aperture with no additional ERF filters will collect $1000 \times 0.1 \times 0.1 \times 3.14/4 = 7.85$ W at the slit.

Many options are available for the slit plate design. The simplest solution is to use a commercial slit plate. Unfortunately the largest market for the slit plates is with spectrographs and monochromators used in commercial, university research, and medical applications. The normal slit gap used is <3 mm long. These are readily available from scientific suppliers like Edmund Optics and Thorlabs. These companies can supply special adjustable slit assemblies normally used for laboratory work, these can have slit lengths up to 8 mm but are relatively expensive (>$250). On the other hand Surplus Shed sell a 6 mm long adjustable slit assembly for less than $20 which, with some attention, can be used successfully. See Appendix C for suppliers; alternatively a DIY construction can be used.

Two single-edged razor blades mounted in a suitable holder and aligned parallel work very well. The bevel edged blades from a pencil sharpener are even better. Simple mechanisms can be used to vary the slit gap (Fig. 6.13).

Reducing Transversalium: Cleaning the Slit Gap

Any dust or imperfections in the entrance slit jaws will cause transversa-lium—dark lines along the spectral image. They can be very difficult to remove from the final SHG mosaic image where they show as a series of lines across the image. The best solution is to remove them at source by cleaning the slit gap. The slit should be opened to around 1 mm and the slit blown with pressurised air. A camel hair brush can also be used. If all else fails then gently rubbing the length of the slit with a wooden toothpick generally gives improved results.

Remaining transversalium needs to be addressed by the application of flats during the final processing (see Sect. 8.4).

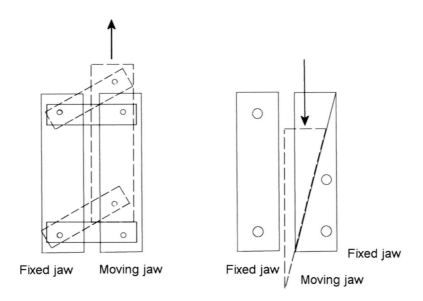

Fig. 6.13 Adjustable slit designs

In Sidgwick's design (Fig. 6.14) parts A and B form the support for the slit jaws. Spacer F holds the securing plate E forming a retaining track for the slit jaws. The bottom jaw C is fixed and the moving jaw D adjusted using the nut and screw H.

The center of the entrance slit gap should be placed on the optical axis, and the gap length aligned with the grooves in the grating. The flat face of the slit plate should face the incoming beam from the telescope to reduce reflections in the spectrograph (see item D in Fig. 6.14).

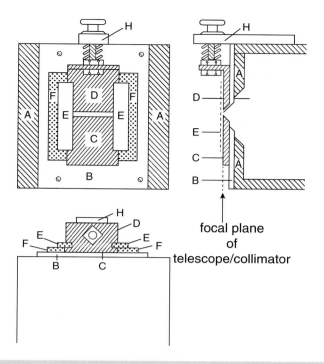

Fig. 6.14 Sidgwick's adjustable slit design (J.B. Sidgwick)

Many amateurs prefer the flexibility of the adjustable slit designs, but a fixed slit gap will work just as well. Commercial laser cut or etched fixed slit plates are available in slit lengths around 3 mm and have 10 µm gaps upwards which may suit the smaller digital SHG.

6.4.1   Setting of the Slit Gap

This is a critical feature of the digital SHG. The width of the gap will define the resolution of the final image, and must be selected to suit the imaging camera pixel size.

The average solar disk is 1920 arc seconds diameter. The plate scale at the slit, say for a 1000 mm focal length is 205 arc seconds/mm (0.2 arc second/µm) giving a solar image diameter of 9.7 mm.

Consider, if we used an adjustable slit 10 mm long and the gap was opened up to 3 mm, with the camera placed behind the slit, we would record a section of the Sun, full height (9.7 mm) by 3 mm wide. This would cover a width 1/3 of the disk. A sequence of three such images would allow us to stack them in a mosaic to give full coverage of the solar disk. The detail recorded WITHIN each image would be still be limited by the seeing conditions, approximately 2 arc seconds.

Going to the other extreme, we could close the gap down to say, 10 μm (giving 2 arc seconds image width) and by taking $(1920/2)=960$ images would, when combined in a mosaic produce the same final image.

How far can we go in reducing the gap even further? Based on the plate scale being discussed, a minimum slit gap of 5 μm is needed to achieve a 1 arc second surface resolution. Unfortunately this is unrealistic. We need to consider the ability of the camera to record this detail. The commonly used Shannon–Nyquist sampling theory says data from at least 2/3 pixels are required to record the minimum resolution.

Based on a camera pixel of say 4.5 μm, the minimum recommended slit gap would become 10 μm. This would give a surface resolution of only 2 arc seconds.

To balance these two extremes, a longer focal length telescope would be required. However that may limit the extent of the solar disk height being recorded. Building up a grid of mosaics would then be necessary to capture the whole disk.

6.4.2   Measuring the Slit Gap

The actual physical slit gap can be easily measured using a laser pointer. The small red laser pointers operate at 6500 Å, the green version can be assumed to be at 5320 Å.

Figure 6.15 shows the set-up required. A handheld laser pointer is directed at and through the slit gap and illuminates a nearby wall/card target, at distance D (mm). The diffracted image caused by the interference at the slit gap will show on the target as a series of bright slightly elongated spots. The size of the slit gap (μm) (S) can be calculated by measuring the distance (X) (in mm) between the centers of the two first order images.

$$S = \lambda_L \times 3 \times D / 10{,}000 \times X$$

Example $D = 500\,\text{mm}, \ X = 45\,\text{mm}, \ \lambda_L = 6500\,\text{Å},$

$S = 6500 \times 3 \times 500 / 10{,}000 \times 45 = 21\,\mu\text{m}.$

6.4.3   Length of the Slit Gap: Long Slits

In stellar spectroscopes the target image is usually positioned very close to the optical center of the entrance slit gap, on the optical axis. In the case of the SHG we are working with a solar image, the size of which increases with telescope

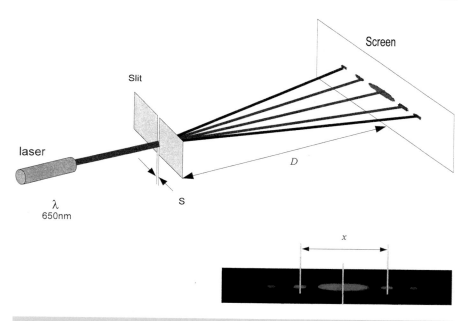

Fig. 6.15 Laser measurement of slit gap

focal length. The effective length of the entrance slit is usually the limiting factor in obtaining full disk images. Slit lengths of 8 mm are not uncommon and this needs to be taken into account with the spectrograph optical design to minimize vignetting and aberrations.

Consider a slit length of 8 mm, and a collimator with a focal length of 200 mm working with an f10 telescope. The normal design parameters would indicate a collimator diameter of $(200/10) = 20$ mm and a matching grating of very similar size.

With an extended entrance slit, called a long slit by the professionals, the light from the top and bottom portions of the slit gap will be off-axis, in this example by up to 4 mm and present an angle called the field angle to the collimator of Tan^{-1} $(4/200) = \pm 1.1°$. This means that the size of the collimator lens (and grating) must be increased to avoid unnecessary vignetting by at least 2×4 mm, giving a minimum 28 mm diameter, 40 % larger. The inclination of the light path from the extremes of the slit length will result in aberrations at the grating (see Sect. 6.6 for more).

6.4.4 Field Lens Compensation

The optical effects of a long slit can be reduced by adding a Field or Fresnel lens immediately behind the entrance slit. This lens refocuses the edge of the beam back towards the optical axis and gives a more even field illumination. Designing an accurate field lens is complex, but Bartolick has come up with a rule of thumb which can be applied to Classical and Littrow spectrograph designs.

The field lens diameter should be larger than the long slit gap, and the focal length $F_{fl.}$ would then be:

$$F_{fl} = 1 / \left(\left(1 / F_t \right) + \left(1 / F_c \right) \right)$$

Example

$$Telescope\ focal\ length, \quad F_t = 1200\,mm$$
$$Collimator\ focal\ length, \quad F_c = 200\,mm$$
$$Field\ lens\ focal\ length, \quad F_{fl} = 1 / \left(\left(1 / 1200 \right) + \left(1 / 200 \right) \right)$$
$$= 1 / \left(0.00083 + 0.005 \right)$$
$$= 1 / \left(0.00583 \right) = 172\,mm$$

This field lens would then allow a smaller more standard sized collimator/grating to be used.

6.4.5   Summary: Entrance Slit

– Use long slit gaps for maximum solar disk coverage
– Match the minimum slit gap to the camera pixel resolution
– Keep the slit gaps clean

6.5   The Collimator

In astronomical spectrographs it is usually stated that the f ratio of the collimating lens needs to match the f ratio to the telescope being used. An f10 telescope will require an f10 collimator. This is generally correct of stellar spectroscopy where the target star is positioned central to the entrance slit. Conventional spectrographs usually restrict the length of the slit to 3 mm to minimise the off-axis effects. In the case of the spectrograph used in the SHG we must face up to the fact that the solar disk subtends a significant angle (±0.25°) and that a long slit is required (>3 mm, usually 6–10 mm).

The diameter of the collimating lens therefore needs to be large enough to accommodate the long slits used in the SHG and fully cover the size of the grating being used. The field angle if not corrected by a Fresnel or field lens requires that the collimator lens be slightly over sized, a bit like using larger secondary mirrors in a Newtonian to increase the FOV. The focal length of the collimator will remain the same as the basic calculations, but the clear lens diameter larger—this then means that the actual design f ratio of the collimating lens will be slightly faster than the telescope.

The emerging beam from the collimator to the grating will also show the field angle, it will be close to a collimated beam but not all parts of the beam are parallel. The same effect gives us the exit pupil and eye relief position when an eyepiece used in a telescope. The larger overall beam size also affects the optimum size of the grating. This again needs to be slightly larger than the usual calculations would indicate.

We usually apply the height of the solar image at the entrance slit to the collimator and grating dimensions to find an acceptable compromise. A slight loss of throughput efficiency is not a major issue when observing the sun; there's plenty of light available. The collimator lens is positioned at its focal length behind the slit gap to provide a collimated light beam to the grating.

Achromatic doublets are used in smaller digital SHG's. Larger instruments can use mirror optics (Littrow and Ebert–Fastie or Czerny–Turner designs). The focal length of the collimator will then be approximately equal to the grating width times the f ratio. Assume a 50 mm square grating and an f5 telescope. This would require a collimator with a minimum focal length of 250 mm. There can be some loss of efficiency if the collimator diameter and grating size does not fully accommodate the long slit gaps used to produce large solar images. This is considered and taken into account in the calculations used in the SimSpec SHG spreadsheet (see Appendix B).

Bear in mind that the imaging system may also influence the final choice of collimator due to the size of the pixels and overall size of CCD chip used.

All achromatic lenses suffer from chromatic aberrations. This means there will be a different focal point for blue, green and red light. The difference in focal length can be greater than 1 % i.e. a lens of 400 mm nominal focal length can show a focal variation of up to 4 or 5 mm. This variation needs to be taken into account when finalising the mechanical design of the spectrograph and means of re-focusing the lens must be included.

Chromatic aberration is not a problem when reflective collimators (mirrors) are used; all wavelengths come to the same focal point.

6.5.1   Precise Focusing of the Collimator to the Entrance Slit

To ensure the best performance from the grating, the collimator must be positioned exactly at the correct focal distance. This is normally confirmed by removing the grating and the imaging system, if necessary, from the spectrograph and viewing the slit gap through the collimating lens, on axis, from the grating location.

A telelens or astronomical finder pre focused to infinity is ideal. When a narrow slit gap is viewed/imaged through this set-up it should be seen in tight clear focus. Adjust the collimator/slit plate spacing as necessary. This will ensure that the resulting beam exiting from the collimator to the grating is well collimated and parallel.

6.5.2   Summary: Collimator

– Collimator focal ratio must match the telescope
– Add a focussing mechanism to refractive collimators
– Critically focus the collimator to the slit gap

6.6   Gratings

The purpose of the grating is to break down the incoming light from the collimator into a spectrum, this is the diffraction of the grating. This is achieved using a series of very fine grooves in the surface of the grating. The number of lines or grooves per mm (l/mm) is a key parameter for all gratings.

More illuminated lines on the grating will give more theoretical resolution; hence use the largest grating you can to get maximum theoretical resolution. Also, the angular dispersion, that is the spread of the spectral image, will be influenced by the l/mm.

The grating surface acts like a mirror and reflects some of the incoming light to form a zero order image. The grooves on the surface then diffract the remaining light into a series of spectra on either side of the zero order image (see Fig. 4.17). The brightest spectrum closest to the zero order image is the first order spectrum, the next brightest the second order and so on.

The light distribution across the zero order image and subsequent spectra is approximately 50 % in the zero order image, 15 % in each of the first order, and 5 % in the second. This is very inefficient and to improve this distribution most professional gratings are now blazed. Blazing is where the shape of the groove in the grating is tilted to preferentially deviate the maximum amount of light possible into one of the first order spectra. This can dramatically improve the efficiency and up to 70 % of the incoming light can be directed into the first order. The direction of the blazed first order spectrum is always marked on the edge of the grating with an arrow.

The efficiency of the grating and the positive effects of blazing can be seen by comparing the brightness of the zero order images produced. Gratings which produce the brighter zero order images are far less efficient.

Gratings come in different sizes and are constructed from a grooved substrate mounted on a glass backing and are available in various l/mm, typically 600, 1200, 1800 and 2400 l/mm. The dispersion for a 1200 l/mm grating will be twice that of the 600 l/mm, so the spectrum formed will be longer, fainter, but the possible resolution higher.

Standard sizes for gratings are 12.5×12.5, 12.5×25, 25×25, 25×50, 30×30 and 50×50 mm. The price increases exponentially with size, and most amateurs use either the 30×30 mm or the 50×50 mm gratings. Gratings can be manufactured as transmission gratings or reflection gratings. Reflection, blazed gratings are generally used in the digital SHG design. To get maximum resolution, a large dispersion (Å/mm) is needed.

This high dispersion can be obtained by either using a high l/mm (1200/1800/2400) or imaging the spectrum in higher orders (second). Note that 1800 l/mm gratings and above when used in the second order will only allow the imaging in the UV blue region of the spectrum. The free spectral range (FSR) of a grating is defined as the wavelength span of the non-overlapped spectrum. In the case of a first order spectrum this is usually quoted as 3500–7000 Å. Above 7000 Å the second order spectrum of wavelengths above 3500 Å will start to overlap the first order wavelengths. This can be seen in Fig. 6.17.

6.6.1 Use of Prisms as Disperser

Replica reflection gratings came into common amateur use in the mid 1900s. Prior to that, prisms were used to provide the dispersion in the spectrograph. The dispersion of a prism depends on the type of glass, the physical size and shape of the prism and the wavelength of the light. The dispersion, unlike a grating is non linear with the highest dispersion in the blue wavelengths and the lowest towards the red end of the spectrum. There is no zero order image or multiple spectra produced by a prism; only one bright unambiguous spectrum is produced.

The refractive index (n) of the glass causes a deviation of the light beam at both the entry and exit points of the prism. Snellius (1580–1626) was the first to investigate this property and developed Snell's law:

$$n = \sin i / \sin r$$

Where i is the angle of the incident beam to the surface and r the angle of refraction from the surface. The refractive index n also varies with the type of glass (crown, flint etc.) and wavelength. The larger the n value the larger the deviation.

For minimum deviation the light should pass through the prism parallel to the base. The incident and exit angles will be approximately half the apex angle of the prism.

The resolution of a prism is given by:

$$R = \Delta n / \Delta \lambda \times \text{base length of the prism (s)}$$

The term $\Delta n / \Delta \lambda$ is the index of refraction—a measure of the prism material dispersion and can vary from 0.013 to 0.0036 (crown glass at 4000 and 6000 Å) to 0.051–0.01 (flint glass at similar wavelengths). A 60° flint prism with a base length of 30 mm would therefore give a resolution R = 1530 at 4000 Å and only R = 300 at 6000 Å. To increase the dispersion and hence the resolution it was not uncommon to see trains of up to seven prisms used in early astronomical spectrographs.

Hale used a 4″, 568 l/mm Rowland grating in his initial Kenwood SHG (1892), but changed to two 60° prisms in his larger Rumford design (1902). He felt that the prisms gave less scattered light and a brighter spectrum than the early Rowland grating.

One downside to the use of prism dispersers is that the use of long slits causes curvature in the recorded absorption lines. The field angle effect at the extremes of the slit length causes the light path through the prism to be longer and the deviation greater. This results in the absorption line appearing as an arc, with the ends curving towards the blue side of the spectrum. The radius of this curvature is approximately equal to the focal length of the imaging lens.

The quality and reasonable cost of the replica reflective gratings as well as the ease of mounting and aligning the grating optical elements has made the use of prisms almost obsolete in the construction of the SHG. For the dedicated amateur who is up for the challenge and wants to replicate the earlier work of Hale etc. good 60° flint prisms can still be obtained from suppliers like ThorLabs and Edmund Optics.

6.6.2   Abbe Number

You will see in the optical catalogues that the type of glass used in the manufacture of quality prisms is quoted in terms of their Abbe number (V_d). This classification was developed by Ernst Abbe (1840–1905) a professor at Jena, Germany, who latter worked for the optical company Carl Zeiss Jena.

The Abbe number, V_d, is calculated by:

$$V_d = \left(n_d - 1\right) / \left(n_F - n_C\right)$$

Where n_d, n_F, and n_C are the indices of refraction for the Fraunhofer helium D-line (5876 Å), the hydrogen F-line (4861 Å), and the hydrogen C-line (6563 Å).

A lower Abbe number indicates more dispersion. Glass with an Abbe number <50 is classified as a flint glass, >50 a crown glass.

6.6.3   Transmission Gratings and Grisms

Although not favoured by many of the SHG builders, a suitable transmission grating can in fact be used. A good example is shown in Sect. 9.11. The best available transmission gratings appear to be 1200 l/mm and can be used independently or in conjunction with a prism or grism system. The grism arrangement is a transmission grating mounted on to, or very close to, a suitable prism or series of prisms, where the total deviation angle of the prism(s) selected brings the target wavelength back onto the optical axis.

The deviation angle for the Hα wavelength with a 1200 l/mm grating is calculated in Sect. 6.6.4 to be 51.9°. We would therefore need a combination of say $3 \times 18°$ deviation prisms to achieve the required total correction angle.

Note: The surface of a grating is VERY delicate and must never be touched at any time. Any fingerprints or marks cannot be removed. Any minor dust can be safely blown from the surface using a hand blower (not the mouth!) Store them when not in use in a dust tight container.

6.6.4   Grating Theory

When a collimated monochromatic light beam goes through a pair of fine slits, the action of the light waves causes both diffraction and interference effects. As the light waves exit the slits they interact at varying angles from the slit to generate a series of interference patterns, where the crest of one wave reinforces the other it forms a bright image or no light where the crests oppose each other. This is illustrated in Fig. 6.16. When white light is considered and a large number of slits or grooves in the grating, these interference patterns vary with wavelength and give rise to a series of spectra. A standard grating can produce up to 5 but sometimes many more orders of spectra.

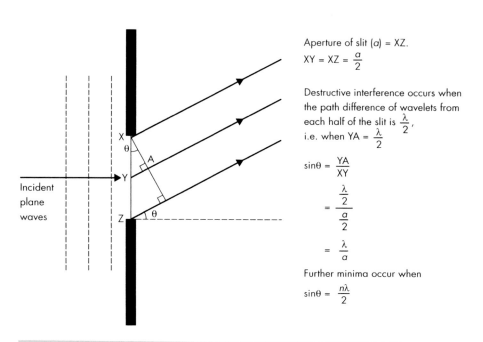

Aperture of slit $(a) = XZ$.

$$XY = XZ = \frac{a}{2}$$

Destructive interference occurs when the path difference of wavelets from each half of the slit is $\frac{\lambda}{2}$, i.e. when $YA = \frac{\lambda}{2}$

$$\sin\theta = \frac{YA}{XY}$$

$$= \frac{\frac{\lambda}{2}}{\frac{a}{2}}$$

$$= \frac{\lambda}{a}$$

Further minima occur when

$$\sin\theta = \frac{n\lambda}{2}$$

Fig. 6.16 Grating –Path length difference (S. Tonkin)

The relationship between the angle of the incident beam (α), the angle of the exit or reflected beam (β), the spectrum order (n), the grating l/mm (N) and the wavelength of the light (λ) are defined by the grating equation:

$$nN\lambda = \sin\alpha + \sin\beta$$

The total angle (Ψ) between the incident and dispersed beam is:

$$\Psi = \alpha + \beta$$

The angular dispersion of a grating is almost linear; the second order spectrum being twice the length of the first order and can over lap on the red portion of the first order. The 4000 Å in the second order will be produced at the same point as the 8000 Å in the first order. The non overlapped length of the spectrum defines the Free Spectral Range (FSR) of the grating (see Fig. 6.17).

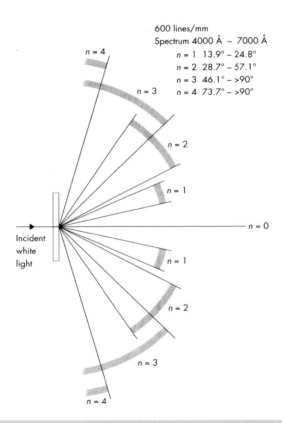

Fig. 6.17 Showing spectra spread from a 600 l/mm transmission grating (S. Tonkin)

The overlap interference in the red portion of the first order spectrum by the blue region of the second order spectrum which limits the FSR can be easily suppressed by using a red filter. A Wratten 25A filter will allow imaging of the Hα (6563 Å) region without second order interference.

For a grating positioned square to the incoming light ($\alpha=0$)

The spectral deviation (β) of the first order spectrum ($n=1$), a 1200 l/mm grating (N) based on 6563 Å (λ) is:

$$Sin\beta = n\,N\lambda$$
$$Sin\beta = 1\times1200\times6563\times10^{-7}$$
$$\beta = 51.9°$$

For CaK, Continuum and Hα light (3934, 5500, and 6563 Å) the first order deviation angle for various gratings is shown in Table 6.2:

Note: Table 6.2 shows that a 2400 l/mm grating will not function with incoming light square to the grating ($\alpha=0$). This grating must be used inclined to the incoming beam ($\alpha>0$). In a Littrow configuration, the grating is inclined (to the optical axis) by ½ the deviation angle (i.e. $\alpha=-\beta$) and the grating rotation required between CaK and Hα wavelengths equal to 0.5Δ(CaK–Hα)

Rotation of grating between CaK and Hα wavelengths:

600 l/mm	4.7°
1200 l/mm	11.8°
2400 l/mm	23.8°

The spectral angular dispersion of a grating is: $d\lambda/d\beta = N\times Cos\beta/n$

For small angles, $cos\beta$ tends to 1, giving the dispersion as $d\lambda/d\beta = N/n$:

Linear dispersion or plate scale $= 10^7\times cos\beta/n\times N\times L$ Å/mm

Where L is the focal length of the imaging lens.

Note: The dispersion of the second order is twice that of the first, a 1200 l/mm grating working in the second order will have the same dispersion as a 2400 l/mm working at first order.

The theoretical spectrograph resolution based on the Rayleigh criteria is:

$$\lambda / \Delta\lambda = n\times N = R$$

Table 6.2 Deviation angles for various wavelengths

Grating	CaK	Continuum	Hα	Δ(CaK-Hα)
600 l/mm	13.6°	19.3°	23.2°	9.4°
1200 l/mm	28.2°	48.7°	51.9°	23.7°
2400 l/mm	56.3°	82.4°	104°	47.7°

Depending on the optical quality of the spectrograph, aberrations and the camera pixel size the actual R value obtained can be 20–50 % of the theoretical. For example, a fully illuminated 50 mm 1200 l/mm grating would probably achieve $R \geq 8000$ compared to a theoretical $R = 60{,}000$.

Anamorphic Factor and Why It Is Important

Sometimes called anamorphic magnification, the anamorphic factor occurs at all wavelengths and is a consequence of the entrance angle α, and the tilt of the grating.

Figure 6.18 shows a monochromatic entrance beam (a) inclined at angle α to the grating normal and the reflected, deviated exit beam (b) from the grating surface at angle β. The total angle between the beams $(\psi) = \alpha + \beta$. The diagram provides the relationship:

$$b / a = \mathrm{Cos}\,\alpha / \mathrm{Cos}\,\beta$$

The ratio b/a is called the anamorphic factor. The grating equation also shows that this ratio will vary with wavelength λ. The ratio will only be 1 (unity) when the grating acts as a mirror (first order image) and when $\alpha = \beta$ in the Littrow configuration.

Figure 6.19 shows the difference in the anamorphic factor when the grating is used at different tilt angles and alignments with the collimator/camera. The grating normal towards the camera is the preferred arrangement.

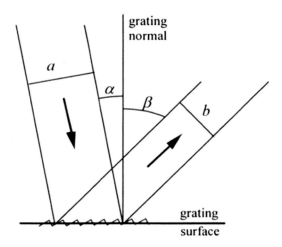

Fig. 6.18 Anamorphic factor

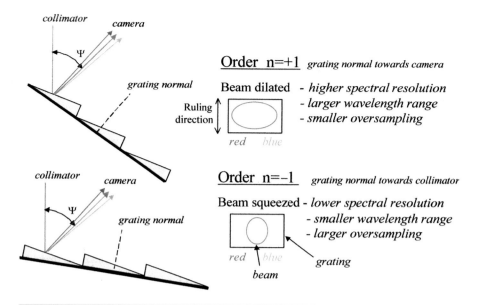

Fig. 6.19 Grating anamorphic factor (J. Allington-Smith)

At ratios greater than or less than 1 the distortion to the beam shape also impacts on the effective width of the slit gap as projected into the camera optics. The anamorphic factor (magnification) compliments the magnification ratio obtained in a spectrograph when differing focal lengths are chosen for the collimator and imaging lenses (see Sect. 6.7.10).

The effective slit width then becomes:

$$S_e = r \times \left(fl_c \,/\, fl_i \right) \times S$$

Where S_e is the effective slit width at the CCD, r the anamorphic factor, fl_c and fl_i the focal lengths of the collimator and imaging lens and S the entrance slit width.

Schweizer (Cerro Tololo Inter-American Observatory) quotes that for a 1200 l/mm grating working at Hβ in second order (tilt angle = 38.45°), the anamorphic factor can allow the use of an entrance slit gap twice the normal gap and still achieve the same resolution. This arrangement also gives increased throughput, which could be very helpful to the amateur.

The anamorphic factor can also lead to a variation of up to 4 % in plate scale across the spectrum due to the 1/Cosβ relationship.

The SimSpec SHG spreadsheet will calculate the anamorphic factor for the chosen spectrograph arrangement (see Appendix B).

6.6.5   Mounting the Grating

The glass grating is very delicate. It should never be touched with the bare hands at any time. Like all glass elements it should never be in direct contact with metal. The best way to attach the grating to the support is by mounted it a small dab of RTV compound on the rear surface. The grating can be mounted onto a small angle support which is capable of being rotated and locked into position (see Fig. 6.20). The direction of the grating grooves should be lying vertically to the base of the holder. All reflective gratings have a preferential blazed direction, this is indicated by an arrow symbol marked on the edge of the grating. This arrow should be point towards the imaging lens for best performance.

Alternatively, to provide a more flexible arrangement which allows the grating holder to accommodate various gratings, the grating can be positioned on the holder and clamped across the top edge to secure it in place. An example of this type of mounting is illustrated in Sect. 9.7. There should be no mechanical stresses applied to the grating.

Note: The center of the grating face should lie on the optical axis, and the axis of rotation must sit through the front surface of the grating, parallel to the direction of the grooves.

If you intend using the SHG across various wavelengths, the most usual case, then the grating needs to be mounted on a rotating platform. This should provide smooth movement through the maximum angle of rotation (see Sect. 6.6.4) and have some means of either calibrating the movement (sine bar/μm adjustment) or

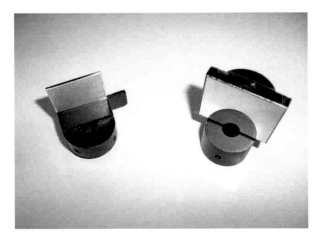

Fig. 6.20 Grating mounts

indicators pre-calibrated to the target wavelengths. Once positioned, the grating assembly is securely locked in position. Various examples of successful grating mounts are illustrated in Chap. 9.

6.6.6 Using the Grating in Second Order and Above

With gratings having 1200 l/mm or less, it is possible to work in the second order and above. The free spectral range allows good access to all visible wavelengths without the need of sort filters. This is achieved by increasing the rotation of the grating to bring the second order etc. deviated beam into the imaging system. The beam width also means that larger gratings are preferred due to their reduction in the front surface dimension. There can also be a reduction in overall efficiency when using smaller gratings. The dispersion in the second order is double the first order. For example, a grating giving 0.8 Å/pixel in the first order will give 0.4 Å/pixel. Similarly the third order would give 0.27 Å/pixel and the fourth order close to 0.2 Å/pixel. The free spectral range (FSR) also reduces with the higher orders used.

This sounds like good news, but unfortunately there are limitations. A 1200 l/mm in the third order will not produce images above 5500 Å, and 4100 Å in the fourth order due to the high deviation angles. Higher order spectra are therefore only useful when working in and imaging around the CaK region. Although the dispersion may be significantly increased, this is not always reflected in dramatic improvements in the actual resolution obtained.

Field Angle (γ Angle) Aberrations

Gratings are designed to function at maximum efficiency when presented with a parallel collimated on axis light beam. This gives a constant entrance angle (α). Off axis slits can be accommodated by retuning or rotating the grating slightly to compensate for the change in angle. However, with long slits, the field angle is at right angles to the entrance angle and cannot be easily changed. This angle, γ (gamma), causes aberrations and can result in slanted and curved spectral lines. The larger the γ angle the larger the effects.

Energy Density Limits

There may be some concern about using expensive gratings in an SHG application viewing the Sun. Could there be thermal/energy damage?

In Sect. 6.4 we show the total incident energy for a typical SHG (100 mm aperture) is less than 8 W. According to Newport Corp., one of the largest grating manufacturers, a "standard replica grating on a glass substrate" has a damage threshold of 80 W/cm². This result shows in our application there is no energy threat to the grating.

6.6.7   Summary: Gratings

- Use large reflection gratings
- Use high l/mm for best resolution
- Blazed gratings give best performance

6.7   The Imaging System

To record the solar spectrum presented by the grating and provide a video (AVI) file, suitable for processing, we need to add a lens and camera combination. The final size and resolution of the solar image will be dependent on the size and focal length of the imaging lens, the CCD frame size and the pixel size of the camera.

The image of the spectrum produced by the grating is formed by the imaging lens. This lens can be another achromatic doublet or the same doublet as the collimator when the Littrow layout is used. The diameter of the lens needs to be large enough to capture the dispersed spectral beam from the grating.

If the collimator/imaging lenses have the same focal length, there is no magnification effect. A 9 mm solar image on the entrance slit is recorded as a 9 mm high spectral image.

To get the height of the solar spectral image to better fit the size of the CCD, a smaller focal length imaging lens can be used. Using a collimator with a 200 mm focal length and an imaging lens of 100 mm focal length will result in a spectral image height of $100/200 = \times 0.5$ the slit image height. The 9 mm solar image example would result in a 4.5 mm high spectral image.

Nothing comes free. Reducing the spectral image will also affect the linear dispersion and resolution. Halving the image lens/collimator focal length ratio will half the dispersion e.g. a 0.7 Å/pixel dispersion would become 1.4 Å/pixel. Also, the entrance slit gap will be projected with the same reduction factor; a 10 μm entrance slit gap would then appear as an effective 5 μm gap at the camera. This may result in under sampling and a loss of resolution if the pixel size of the camera is too large. For the smaller digital SHG designs standard commercial photographic lenses work well. There are many good to excellent old film camera lenses available in the second hand market. The Olympus Zuiko lenses make a good choice as do the wide range of M42 Pentax fitment lenses. DIY adaptors can easily be made to allow the mounting of the CCD camera to any of these lenses. One added benefit of using a camera lens is that you get a focusing mechanism for free.

For Classical and Littrow spectrograph designs, the front aperture of the imaging lens should be positioned as close to the grating as possible, without interfering with the overall lightpath. This reduces the aperture required and can give more flexibility to the choice of lens. If you intend to work in second order or above, the required red sort photographic filter can be mounted on the front of the imaging lens.

The imaging lens if properly matched to the grating and collimator lens should work at close to infinity focus. This then confirms the grating is working in the optimum collimated beam. The SimSpec spreadsheet in Appendix B allows the design of any lens combination to be checked and verified.

6.7.1   Imaging Camera

The key to success with the digital SHG is the choice of recording camera.

A fast frame mono camera capable of producing an AVI file is the minimum requirement. Cameras like the Philips webcam, TIS DMK series (DMK21/31/41/51), the ASI120mm and the Point Grey Grasshopper3 (USB3 only) have been used. The trade off is the frame size vs. frame rates.

Mono cameras have better pixel resolution than the equivalent color camera. The video capture setting available in some of the latest DSLR's is not capable of meeting the needs of the digital SHG and is not recommended.

The main characteristics of the CCD chip which impact on their effective use in the digital SHG are:

* Sensor (chip) size
* Pixel size
* Frame rate (fps)

6.7.2   CCD Chip Size

The physical useful sizes of the sensor in the typical amateur camera vary from 3.6 mm × 2.7 mm in a typical webcam to 11.3 mm × 7.1 mm (ASI 174). When used with different dispersion/plate scales, obviously the larger CCD will allow more of the solar disk corresponding to the height of spectral band and a greater extent of the spectrum wavelength coverage to be recorded. Some popular CCD chips are compared in Table. 6.3. It should also be noted that some of the latest cameras offer a Region of Interest (ROI) feature which can significantly improve the frame rate, discussed later this chapter.

The size of the chip determines how much of the solar disk can be recorded in each exposure. The smaller chip webcams type camera of 2.7 × 3.6 mm, if used with shorter focal length imaging lenses to record the full disk, will limit the resolution obtained. Assembling a series of exposures into a full disk image mosaic is therefore the way to go.

Larger chips like the DMK51 and ASI 174 can give better results, but can also suffer lower frame rates (fps). Using the larger chip dimension, the size of the solar disk recorded (as a spectral band) would be limited to a height of 6 or 7 mm.

Table 6.3 Typical CCD camera characteristics

Camera	CCD chip	Frame rate (fps)	Pixel size (μm)	Pixel array	Chip size (mm × mm)	Resolution (μm)
Webcam	ICX098BQ	60	5.6	640 × 480	3.6 × 2.7	11.2
DMK618	ICX618ALA	60	5.6	640 × 480	3.6 × 2.7	11.2
ASI 120	MT9M034	60	3.75	1280 × 960	4.8 × 3.6	7.5
Flea3	ICX424	80	7.4	648 × 488	4.9 × 3.6	14.8
DMK41	ICX205AL	15	4.65	1280 × 960	6.0 × 4.5	9.3
ASI 130	MT9M001	30	5.2	1280 × 1024	6.6 × 5.3	10.4
DMK51	ICX274AL	12	4.4	1600 × 1200	7.0 × 5.3	8.8
G'hopper3	ICX687	15	3.69	1928 × 1446	7.1 × 5.3	7.4
ASI 174	IMX174LLJ	128	5.86	1936 × 1216	11.3 × 7.1	11.7

As an example, consider a digital SHG where the telescope is 1000 mm focal length, producing a solar image 9.4 mm diameter on the entrance slit, a collimator of 200 mm focal length and an imaging lens of 200 mm. An AVI file recorded by a webcam type chip would give a coverage of 3.6/9.4=0.38 the solar diameter, some 12.16 arc min (730 arc seconds). This would then mean building up at least three images across the solar surface; one across each of the polar region and one for the central region and assembling the images into a mosaic to present a full disk image.

In this example, an alternative would be to reduce the focal length of the imaging lens to <80 mm, say 75 mm, resulting in a reduction magnification (75/200) of ×0.375. The 9.4 mm solar image on the entrance slit would then give a spectral height of (9.4 × 0.375) 3.5 mm on the CCD. The full solar disk then fits the available height of the webcam chip. The downside to this method is the loss of resolution. The height of the disk will only be 640 pixels high, a plate scale of 2.9 arc seconds/pixel, with a probable resolution of 6 arc seconds.

The impact of negative magnification on the spectral resolution will be discussed in the next section.

6.7.3   Pixel Size

The final resolution of the digital SHG will depend on the pixel size of the CCD camera. The Nyquist sampling criteria defines a minimum of a 2–3 pixel sample to define the effective resolution. This is shown in Table 6.3.

It also means that the effective slit gap as projected onto the CCD should be close to this size. Generally this means ideally working with an entrance slit gap of 10–20 μm.

Due to the Bayer matrix (about which more later) in one shot color (OSC) chips, a minimum of four pixels need to be illuminated to record all the incoming wavelengths of light. The final resolution will be slightly less than a mono CCD chip and will depend on the Debayering model used. The effective slit gap width means the projected size of the physical entrance slit. If the collimating lens and camera lens have the same focal length then there is no magnifying or reducing effect. There is no benefit in having the camera lens focal length longer than the collimator. However, to achieve maximum coverage of the solar disk, it is sometimes useful to reduce the camera focal length.

We also need to take into account the effect on the slit gap, which will also be reduced. Using the same parameters as the example in Sect. 6.7.2, an imaging lens of 75 mm would mean a reduction of ×0.375. In order to maintain an acceptable sampling ratio, the effective slit gap at the CCD must not be less than 10 μm, meaning the entrance slit gap would then need to be increased to $(10/0.375)=27$ μm!

6.7.4   Bayer Matrix

In all color cameras the color response of the individual pixels are controlled by the use of a colored filter array (CFA), called the Bayer filter. This consists of a series of red, green and blue filters (RGB) or cyan, magenta, and yellow (CMY) filters placed in a matrix over the CCD pixels. Usually these are 50 % green, 25 % red and 25 % blue (Fig. 6.21). Most one shot color cameras use the RGB filtering system. Effectively each pixel records a monochromatic intensity signal of the incident light.

To achieve a color image output from the various pixels the Bayer matrix must be translated back into a RGB color. Various debayering algorithms have been developed. Some, like the Bilinear method, give low resolution compared with the more sophisticated VNG (Variable Number of Gradients) model. The positioning of the spectral image on the Bayer matrix can also significantly affect the resolution and response of the CCD.

It is recommended therefore that the spectral image always be positioned as close to the horizontal or vertical axis of the chip as possible.

6.7.5   Surface (Spatial) Resolution

The image size and pixel size will determine plate scale (arc second/pixel). In the previous example using 200 mm lenses, the extent of the solar surface being recorded was 730 arc seconds. This was on a chip with 640×5.6 μm pixels. The resulting scale would then be $730/640=1.14$ arc second/pixel. Due to minimum sampling requirements the surface resolution would be at least twice this figure, say 2.5 arc seconds or approximately $736 \times 2.5=1840$ km. Any features like granules, filaments, or plage that are less than this physical size would not be clearly seen on the solar surface. Luckily, most of the interesting features we wish to record are

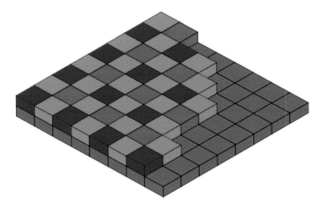

Fig. 6.21 Bayer matrix (WIKI)

larger. Filaments can have sizes up to 10,000 km wide by 100,000 km long and prominences can extend >50,000 km above the surface. The final spatial resolution unfortunately will always be influenced by the local seeing conditions, and may be much worse than the above figures would indicate. Notwithstanding that, it's always good to know what could be achieved under excellent conditions—when they occur.

6.7.6 Quantum Efficiency

At this point in regular imaging one would normally discuss the sensitivity and quantum efficiency of the CCD camera. With the amount of light available for solar imaging, however, this is not as important a factor as it is for night-time spectroscopy.

The quantum efficiency (QE) graph, sometimes called the Sensitivity Curve, for the CCD chip shows the response across the spectrum. This curve usually peaks around 5500 Å. For the imaging of solar spectra, the efficiency only really has importance when working in and around the CaK line. You will see from the QE curves illustrated (Figs. 6.22 and 6.23) that the response of all the common silicon based CCDs drops off quickly in the UV. Efficiencies of 20–30 % below 4000 Å are not uncommon. The silicon construction of the CCD also means that the response curve tends to zero around 11,000 Å. Special IR detectors are required for imaging this far into the red region of the solar spectrum. Note that most of the supplier literature shows these curves as relative response. These can be misleading inasmuch as they do not give an actual QE figure but rather a percentage of the maximum. Most chips will have a maximum absolute quantum efficiency below 60 %.

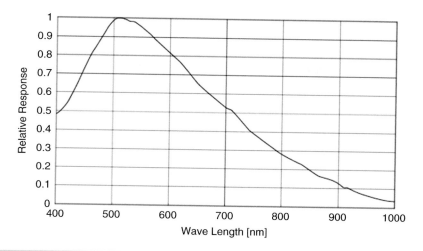

Fig. 6.22 QE curve for ICX098BL (DMK) sensor (Sony)

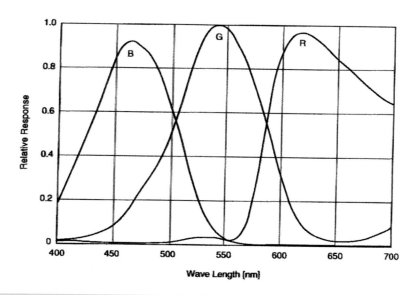

Fig. 6.23 QE curve for ICX098BQ (Color webcam) (Sony)

In the case of one-shot color chips (Fig. 6.23) the RGB response curves can show significant dips in efficiency between the various color filters used in the Bayer matrix.

6.7.7   Spectral Dispersion and Resolution

Again the combination of effective slit gap, imaging lens focal length, the camera pixel size, the size of the grating, and the l/mm will determine the spectral resolution in the final image. Table. 6.4 shows some typical results.

Note: The linear dispersion (Å/pixel) is not the same as the resolution (Å) obtained from the system.

From this table it can be seen that the best resolution is for any nominated pixel size obtained from a combination of high l/mm (2400 l/mm), and longer imaging lens focal length (200 mm). The resolution results shown for the 2400 l/mm grating (0.28 Å) would give the possibility of extracting a spectroheliogram with a similar bandwidth resolution to that obtained with the best double stacked Hα filter configuration.

For maximum solar surface detail the lowest spectral resolution is better at delivering enhanced contrast. For the recording of prominences, a wider bandwidth can give good results due to their Doppler shifts.

6.7.8   Camera Exposure Times

Generally there is a relationship between the exposure and the frame rate. Obviously the individual frame exposure can't be greater than the frame rate interval.

Table 6.4 Dispersion and resolution for various gratings

Grating (1/mm)	Imaging lens (fl, mm)	Pixel size (μm)	Dispersion (Å/pixel)	Resolution (Å)
1200	100	5.6	0.43	1.3
	200	5.6	0.21	0.84
1800	100	5.6	0.25	0.79
	200	5.6	0.13	0.49
2400	100	5.6	0.14	0.45
	200	5.6	0.07	0.28

Note: Most of the analysis software available to the amateur calculates wavelengths in Angstroms. This is easily converted, if needed, to nanometres (nm) by dividing the Å value by 10 i.e. 1000Å = 100 nm

If the selected exposure is 1/60 s, then the best frame rate would be 50 frames/s, or even less, allowing for the electronic transfer of the image to the storage medium. For conventional filter solar imaging this usually doesn't become a problem. The exposure time is usually selected as short as possible to freeze the moments of best seeing. By stacking only the best quality frames from the AVI file, the subsequent processing can result in a very sharp final image.

With the digital SHG only one complete spectroheliogram image can be processed from any one AVI spectrum. There is always the option of preparing multiple images with varying bandwidth selection, but the underlying detail can only show what was recorded during the AVI file. Each frame of the AVI contains effectively a separate image strip of the solar surface. Any rapid solar feature changes during the AVI exposure, like those in and around active areas and prominence movements, may result in the loss of definition. Unfortunately the ability to stack selected good individual frames is not a real option at the moment with the digital SHG. The changing seeing conditions during the AVI exposure have the same effect. One solution is to work at faster scanning rates and thereby reduce the total time taken to collect the AVI file. This is discussed in detail in the next section.

Finally, the minimum exposure selected should be sufficient to record the full depth of the target absorption line. The deeper we can go into the line, the better the contrast and the eventual recording of details in the various solar features. Longer exposures can mean that the surrounding continuum may be over exposed. This has no impact on the final spectroheliogram produced.

6.7.9   Frame Rates/Scanning Rates

The last piece of the puzzle is working with the frame rates of the camera and balancing this with the scan rate of the solar image across the entrance slit. Consider a camera with a frame rate of 20 fps (frames per second). If the solar image is allowed to trail (no drives) across the slit, then we would obtain $20 \times 120 = 2400$ images. Each image would be recorded 1/20 s after the previous. During this time the Sun would have moved $1920/120 \times 1/20 = 0.8$ arc second

The entrance slit gap, say 10 μm on a telescope of 1000 mm focal length will record an image strip with a width of 2 arc seconds. This then gives a series of exposures, each covering 2 arc seconds, which will be superimposed on a 0.8 arc second region of the solar disk during processing. This over sampling is very common with digital SHG designs. It also infers that the sampling frame rate could be reduced to 10 fps with minimal loss of detail.

Following through with the same example, using a 1200 l/mm grating, 200 mm focal length imaging lens and 5.6 μm pixel camera from Table 6.4, the dispersion will be 0.21 Å/pixel and a probable resolution of around 0.84 Å. If we extract a strip from the spectral image based on the Hα wavelength of 2/3 pixel width, the detail recorded will be at a 0.84 Å resolution, but will also cover the same region of the solar surface recorded by the entrance slit—2 arc seconds wide.

After processing, the end result will show the solar surface at a spatial resolution of 2 arc seconds, in the selected wavelength, with a bandwidth of 0.84 Å. Considering the latest range of cameras include a Region of Interest (ROI) option, the opportunities this can give for increased frame rates become exciting.

It will be demonstrated in Sect. 9.13 that by reducing the active image area of the chip (ROI) the frame rate can be significantly increased. A 1200×800 pixel sensor could be re-set to only record a ROI of say 800×100 pixels, the final ROI dimensions being selected to match the solar image size and the extended spectral bandwidth required. This has the impact of increasing the frame rate from 30 to 300 fps, a tenfold increase. With such a high frame rate we could then consider speeding up the image acquisition. This could be done by scanning the solar disk with the RA drive rate increased to ×32 (a speed available on some EQ mounts). At such a speed the scan would take (120/32) 3.75 s and acquire an 1125 (3.75×300) frame AVI file. This in theory would maintain the resolution of the conventional drift scan and effectively freeze any movements of the solar features, the process then almost achieving a live view outcome.

The downside with this solution is the fact that the individual exposures must be reduced as the frame rate increases. With the conventional drift scan set at 30 fps we could set the individual exposure to any value up to approximately 1/30 s. At the higher frame rate of 300 fps the maximum exposure would be limited to less than 1/300 s. With smaller aperture SHGs this may not be a long enough exposure to fully record the depth of the absorption lines, and give a reduced contrast outcome.

The frame rate and scanning rate options are certainly a fruitful area for future experiment and have the potential to significantly change the rules for future SHG imaging. Only time will tell.

6.7.10 Aspect Ratio and Distortion

Continuing the example from Sect. 6.7.2, based on a 1000 mm focal length telescope the solar diameter would present a spectral image with a height of 9.4/1000×5.6 = 1678 pixel, assuming no magnification in the spectrograph.

If the AVI was collected at a frame rate of 1/20 s and normal no drive drift rate of 120 s then we would have 120×20 = 2400 frames in the AVI which would extend from one edge of the solar disk to the other. Extracting the spectroheliogram strip at two pixel wide to achieve the best resolution would result in a series of 2400 images of two pixel wide. The resulting spectroheliogram mosaic would have 2400×2 = 4800 pixels across the width of the image.

This will make an impact on the mosaic image produced. One axis of the solar image will be ×2.7 the size of the other, the height being 1678 pixel and the width 4800 pixel. Any oversampling as per this example, where a two pixel wide spectral strip is being used against one pixel on the solar disk, will cause the mosaic image to appear elliptical.

The amount of the distortion will vary with the pixel count of the strip being used for the bandwidth selection. Obviously with wider bandwidths (0.8 or 1 Å) would result in extreme image aspect ratios. If a bandwidth of only one pixel is used (not recommended) then the aspect ratio would be minimal, only $2400/1678 = \times 1.43$

All this distortion can be easily corrected in the subsequent processing. All of the software (see Chap. 8) used for processing the spectroheliograms demonstrate this capability. If your camera supports Region of Interest settings it may be worthwhile trying a variety of different settings to determine if a smaller ROI can still record the required data. A smaller ROI gives a smaller AVI file outcome and also has the potential to increase the frame rate and give some control over rapid solar movements and seeing conditions.

6.7.11 Bandwidth Selection

The dispersion (Å/pixel) will depend, as we have found, on three major variables — the grating l/mm, the imaging lens focal length and the camera pixel size. As shown in Table 6.4, we will need two or more pixels to record the spectral resolution. A more accurate assessment can be obtained by confirming the FWHM resolution of the SHG using a neon reference lamp (see Sect. 7.4).

The minimum bandwidth strip selected should therefore equal the resolution. This infers that the minimum strip width selected in the AVI to produce the spectroheliogram should be two pixels wide or greater. In the processing software we always have the ability to select and use a one pixel wide strip; this will result in under sampling and loss of detail.

The only way to further improve (reduce) the effective bandwidth is to increase the dispersion and hence the resolution. Selecting a strip wider than the resolution pixel count will increase the recorded bandwidth. Also, the more pixel columns selected in the strip, the larger the aspect ratio to be corrected in the final image.

6.7.12 Wide Slit Recording of Prominences

In the quest to observe the solar chromosphere, the early observers relied on the occasional solar eclipse to hide the bright photosphere and at the moment of totality expose the edge of the solar atmosphere — the chromosphere. The bright hydrogen prominences attracted their attention and although the use of the spectroscope was still in its infancy it was quickly applied to the problem. The prominences could be observed with a slit spectroscope, but just as one narrow strip at a time. Janssen did suggest using a vibrating slit to allow the full extent of the prominences to be observed, though his early experiments were unsuccessful. Huggins fortunately quickly discovered that the prominences were so bright compared with the surrounding background light that they could be observed in all their glory by simply positioning the prominence on the slit and then opening it wider. The slit gap was positioned tangential to the solar disk and scanned around the edge. Once an indication of a prominence was noted, the slit gap was opened up giving a resulting view of a prominence as shown in Fig. 6.24.

This procedure worked so well, that any further development of the vibrating slit idea was put on the back burner for over 20 years.

We can replicate this wide slit method and obtain images of the prominences. The same digital SHG equipment is used, but instead of producing an AVI and separating the wavelength and bandwidth of interest, we can open the slit and record the prominence as we would a normal image. These can be cropped, stacked and sharpened like any other solar image taken by conventional filters.

6.7.13   Summary: Imaging System

– Use mono chipped cameras for maximum resolution
– Use large chipped cameras for maximum spectral coverage
– Use small pixel sensors to maximise resolution
– Use fast frame cameras to minimise seeing
– Match the pixel resolution to the imaging system
– Consider ROI as a means of further improving the frame rate

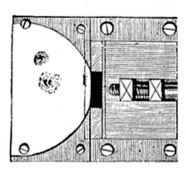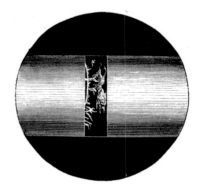

Fig. 6.24 Application of wide slit to view prominences (Young)

Further Reading

Palmer, C.: Diffraction Grating Handbook, Newport Corp. (2005)
Schroeder, D.J.: Astronomical Optics. Academic Press (1987)
Eversberg, T., Vollman, K.: Spectroscopic Instrumentation. Springer (2015)
Nagler, A.: Letters to the Editor, Sky & Telescope, June 2015, p10
Harrison, K.M.: Astronomical Spectroscopy for Amateurs. Springer (2011)
Wodaski, R.: The New CCD Astronomy. New Astronomy Press (2002)
Suiter, H.R.: Star Testing Astronomical Telescopes. Willmann Bell (2001)
Sidgwick, J.B: Amateur Astronomer's Handbook. Enslow Publishers (1980)
Thorn, A., Litzen, U., Johansson, S.: Spectrophysics. Springer (1999)
Tennyson, J.: Astronomical Spectroscopy. Imperial College Press (2005)
Robinson, K.: Spectroscopy: The Key to the Stars. Springer (2007)
Ingalls, A.G., (Ed): Amateur Telescope Making Vol1. Scientific American (1980)
Ingalls, A.G., (Ed): Amateur Telescope Making Vol2. Scientific American (1978)
Ingalls, A.G., (Ed): Amateur Telescope Making Vol3. Scientific American (1977)

Webpages

http://www.cfai.dur.ac.uk/old/projects/dispersion/grating_spectroscopy_theory.pdf
http://articles.adsabs.harvard.edu//full/1979PASP...91..149S/0000150.000.html
http://nickkonidaris.com/2014/10/17/anamorphic-factor/
http://www.baslerweb.com/en/products/area-scan-cameras/ace/aca1600-20gm
http://astrosurf.com/buil/us/spe2/hresol4.htm
http://www.astrosurf.com/buil/
http://spiff.rit.edu/classes/phys301/lectures/comp/comp.html
http://www.thorlabs.com/newgrouppage9.cfm?objectgroup_id=148
http://www.horiba.com/scientific/products/optics-tutorial/diffraction-gratings/
http://refractiveindex.info/index.php?group=SCHOTT&material=N-SF2&wavelength=0.4
http://beloptik.de/en/left/erf-protection-prefilter-energy-rejection-filter/
http://nedwww.ipac.caltech.edu/level5/ASS_Atlas/frames.html
http://spectroscopy.wordpress.com/instruments/
http://www.starrywonders.com/ccdcameraconsiderations.html

Chapter 7

Construction of a Digital SHG: Two Examples

In the previous chapter the various components required for an SHG build were discussed in detail. In this chapter, a couple of design concepts are presented which could be easily constructed by the amateur and then combine these solutions with the option of using two different imaging cameras.

Option 1: A small refractor based SHG based on the Classical layout.
Option 2: A long focal length Littrow design.

Each of these options will then be refined for use with a webcam type camera (DMK21, 5.6 μm pixel, 480 × 640 array, 2.8 × 3.6 mm chip) and one of the large cameras (DMK51, 4.4 μm pixel, 1600 × 1200 array, 5.3 × 7 mm chip) for comparison.

It is recommended before you start on SHG design and construction that you download the SimSpec SHG spreadsheet from the Astronomical Spectroscopy Yahoo group files area. This will allow you to experiment on paper with available lenses and gratings to determine the dispersion outcome and probable end resolution.

After discussing the build, methods to successfully mount and test the SHG are outlined along with information on how to carry out the calibration in wavelength of the spectral image. The chapter concludes by providing step by step instructions on obtaining the necessary solar AVI file used to produce the final spectroheliogram.

© Springer International Publishing Switzerland 2016
K.M. Harrison, *Imaging Sunlight Using a Digital Spectroheliograph*,
The Patrick Moore Practical Astronomy Series, DOI 10.1007/978-3-319-24874-5_7

7.1 Option 1: Small Refractor Classical Spectrograph

Referring back Fig. 6.1, this shows a typical small refractor, Skywatcher ST80 (80 mm aperture, 400 mm focal length, f5) mounted to a classical spectrograph and DMK camera. Let's look closer at this design and how we can assemble such an arrangement. Refer back to the previous sections for background on the individual components and general design requirements.

The 80/400 achromat will produce a solar image at focus of $400/109 = 3.7$ mm diameter. This then tells us that for full disk spectral imaging that the entrance slit needs to be about 4 mm long, and if there is no magnification in the spectrograph, than the final spectral image presented to the CCD will be 3.7 mm in height. From the details of the DMK cameras, this would be too big an image for the DMK21 (3.6 mm maximum frame) but acceptable for the DMK51 (7.0 mm frame size). Let's work with the DMK21 solution. We then need to provide a reduction in magnification between the collimator lens and the imaging lens of at least $3.6/3.7 = \times 0.95$ to fit the solar image onto the chip. In reality, that doesn't allow any flexibility for alignment of the image or obtaining reasonable background coverage for prominences etc. A better solution would be to target say x0.8 reduction. That would give a solar image of 2.96 mm on the 3.6 mm chip.

The telescope is an f5 system, this means that the collimator lens has to also to work at f5. The aperture of the collimator will be defined by the grating used (plus allowance for the solar diameter). The standard available sizes of the commercial gratings mean either a 25, 30 or 50 mm square grating could be used. The choice of the grating l/mm is always to select the highest available. This would be the 1800 l/mm or even better the 2400 L/mm blazed (for 5000 Å) reflective grating from a quality supplier like Optometrics (Optometrics Part # 3-3185 or 3-3242). We will base the design on the 1800 L/mm. If more resolution is required the grating can always be upgraded to a 2400 l/mm later.

The choice of grating size now comes down to a compromise between physical size of the SHG and the possible theoretical resolution of the grating. If a 25 mm square grating is selected we would need a collimator of $25 \times 5 = 125$ mm focal length, a 30 mm would be 150 mm and the 50 mm gives a 250 mm requirement. The collimator focal length adds to the overall length of the SHG. The 400 mm telescope would then have a 125/150/250 mm (plus allowance for the grating assembly) length of spectrograph hanging behind the focus. This has to be well supported and aligned with the telescope optical axis and then be able to be mounted (and balanced) on your mounting. The theoretical resolution of the 30 mm grating would be (30/25), 20 % greater than the 25 mm option and only (30/50) 60 % of the larger grating. Our experience shows that the PSF of the optical system (see Sect. 6.2.1) plays a more important part in influencing the outcome than just assessing the theoretical performance. The choice of size grating is also influenced by the need to find a collimator/imaging lens pair which will not only fully illuminate the grating but provide the required reduction factor of x0.8. That being the case the 30 mm grating would normally be selected at this stage. Possible amendment is discussed later.

We start with the following components:

Telescope—80 mm, 400 mm focal length, f5
Collimator—30 mm, 150 mm focal length, at f5
Grating—30×30 mm, blazed 5000 Å, 1800 l/mm reflective grating

Moving on to consider the imaging system; the reduction magnification required is x0.8, which defines the imaging lens focal length. The lens needs to be 0.8×150=120 mm focal length. The aperture of this lens also needs to match the dispersion beam size from the grating. Checking through the Surplus Shed stock lenses only comes up with a few 30 mm diameter around 172–175 mm focal lengths and nothing around the 120 mm focal length. Thorlabs however show some interesting 50 mm diameter lenses with 180 mm and 150 mm focal lengths (Part # AC508-180-A and AC508-150-A). The reduction ratio 150/180=×0.833 giving a solar image of 3.1 mm would be acceptable. Using a 50 mm lens as a collimator in this f5 application would allow the aperture to be stopped down to 40 mm.

Based on the available lens, we can now review the size of the grating. The collimated beam will now be 40 mm diameter which would overfill the 30 mm grating. A better choice would be to therefore upgrade the grating to the 50 mm size (Optometrics part# 3-5185).

At this stage we now have:

Telescope—80 mm, 400 mm focal length, f5
Collimator—50 mm, 180 mm focal length, working at f5
Grating—50×50 mm, blazed 5000 Å, 1800 L/mm reflective grating

The spacing of the individual lenses/grating in the spectrograph will influence the physical size of the imaging lens required. The classical optical arrangement, as shown in Fig. 6.2.1, gives an indication of the possible problem and constraints. The collimator lens is positioned on the optical axis aligned with the center of the grating. To ensure that there is no vignetting between the lenses this means that the imaging lens must be positioned very close to the collimator but not encroach into the collimated beam. The total angle between the collimator and the imaging lens will probably end up between 35 and 40°. We will work with 35°. The smaller the total angle the better the results. This will give a spacing between the imaging lens and the grating of around 60 mm. The larger this spacing, the larger the size of the imaging lens required. This calculation is again readily available from the SimSpec SHG spreadsheet and this shows ideally a 68 mm diameter lens positioned at 60 mm from the grating would be required. The spreadsheet also shows that when the grating is set for imaging at Hα (6563 Å) the angle (α) of the grating to the collimator will be 55.8° and at this angle a grating width of 64 mm is required to fully cover the collimated incident beam. This is much larger than the 50 mm grating size selected.

With astronomical spectrographs the solution would be obvious: increase the size of the grating or reduce the operating f ratio. A loss of 20 % throughout (50 mm vs. 64 mm) would mean a reduction in the limiting star magnitude by almost half a magnitude. For the solar spectrograph the loss of efficiency is not as critical. We have more than enough input light. Remember that the incident beam will be larger than the grating surface and may cause internal reflections behind the grating which may need to be addressed. The limited size of the grating also acts as an aperture stop to the imaging lens and rather than the calculated design dimension we can safely reduce the diameter to 50 mm.

We need now to look at the entrance slit to compliment the above optical design. Using the DMK21 imaging camera with the 5.6 μm pixel size and working to the Nyquist sampling tells us the effective slit gap should be around 12–15 μm; with a reduction factor of ×0.833 the physical entrance slit gap would then be 15–18 μm. The preferred solution would be to use the "cheap and cheerful" Surplus Shed adjustable slit assembly (Part # M1570D) this provides a slit gap from 0 to 3 mm width and the length of 6 mm. Unfortunately stocks always seem to be limited and they are sometimes not available. Other standard commercial slit sizes are 15 and 20 μm widths and only 3 mm long (Edmund Optics/Lennox laser/Thorlabs). Unless you a willing to spend more than $250 on a slit assembly—Thorlabs Part # VA100. The amateur answer is to make a DIY solution. Section 6.4 details the design variations and options available.

The simplest variant is to use two pencil sharpener blades mounted on a 1.25″ diameter aluminum block with a central hole of say 6 mm to 10 mm using fridge magnet sections to hold the individual blades and setting the gap using the laser pointer method. A similar but slightly different example can be seen the Defourneau's SHG in Sect. 9.7; Slaton also makes use of the magnetic material in his SHS, Sect. 11.2.3.

Means of focusing the solar image onto the entrance slit and focusing the imaging lens onto the CCD need to be finalized. The existing telescope focuser could easily be used for the former. A favorite and very successful method of focusing the CCD itself is to fit a small helical focuser between the imaging lens and the CCD camera. Focusers like the Borg/Hutech 1.25″ non-rotating version, Part # 7315 or the Baader #2458125 focuser provide up to 5 mm focus travel, T2 mounted thread and 1.25″ fitting.

We now have a final design solution:

Telescope—80 mm, 400 mm focal length, f5
Entrance Slit—15–20 μm gap, 6 mm long (pencil sharpener blades)
Collimator—50 mm, 180 mm focal length, working at f5
Grating—50×50 mm, blazed 5000 Å, 1800 L/mm reflective grating
Imaging lens—50 mm, 150 mm focal length, working at f3
Imaging camera—Webcam/DMK21, 5.6 μm pixel, 480×640 array, 2.8×3.7 mm chip.

This configuration, according to the SimSpec SHG spreadsheet, would give a dispersion of 0.15 Å/pixel, a resolution (at Hα –6563 Å) of 0.61 Å (R = 10800) and a wavelength coverage of 244 Å. This performance compares favorably with a standard Hα etalon filter and would certainly show all the chromospheric features.

Before we look at the method of construction and assembly, what would happen if we used the larger DMK51 on this SHG design? The larger chip size (5.3 mm × 7 mm) would allow the same lens (50 mm, 180 mm focal length) to be used for both the collimator and imaging lens as we would not require a reduction magnification. The resulting solar spectrum would then have a full height of 3.7 mm. To maximize the use of the larger DMK51 camera chip, ideally the telescope would have to be replaced with one having a long focal length say, 700 mm focal length which would then give a solar image of 6.4 mm diameter. The larger grating would be retained and the small pixel size (4.4 μm) could allow the entrance slit gap to be reduced to 10–12 μm.

7.1.1 Construction Notes

The spectrograph optics can be assembled into a 70 mm × 70 mm square section housing made from 6 mm plywood. The collimator and imaging lenses are easily mounted in 70 × 70 mm 6 mm plywood partitions with suitable holes drilled/filed to size and fixed in place with three or four dabs of hot melt glue around the circumference. Note that the more convex surface of the lens points towards the grating. The positions of the lenses, grating and slit can be verified by dry assembly, laying them out on a 6 mm thick baseboard before cutting it to size. This process will confirm the best angle which can be achieved between the collimator lens and the imaging lens (this should be close to 35°). A similar shaped board is then used as the lid of the spectrograph housing. Allow at least a 40 mm space in front and behind the grating axis position to give the freedom for free rotation.

The grating can be mounted in a 12 mm × 1.5 mm × 50 mm long aluminum angle frame. Note the direction of the blaze arrow on the edge of the grating — this should point towards the imaging lens. A simple rotating base can be made from a small turntable, a 65 mm disk of 6 mm ply mounted to the base board by a M4 × 20 mm long countersunk machine screw glued into the turntable. The grating angle is then positioned central to the top surface with the front face of the grating sitting over the center of the M4 screw and glued in place. A 4 mm hole drilled through the base, aligned with the intersection of the collimator/imaging lens optical axes provides the axis of rotation for the grating holder. Between the turntable and the base, circular PTFE spacers should be placed, both to reduce friction during adjustment of the grating angle and to lift the center of the grating to the optical axis. These may be cut from plastic milk bottles or something similar. This will need 2.5 mm of spacers. The extended thread of the M4 screw sitting outside the housing can be fitted with a washer an wing nut to lock the position of the grating and can also have a small indicator lever fitted to provide wavelength calibration. See the examples in Chap. 9 for even more ideas.

The entrance slit assembly is positioned on the optical axis at a nominal distance equal to the focal length of the collimator. Before finalizing the slit mounting detail we have to consider how the spectrograph will be mounted to the telescope. If the slit assembly is constructed using a 1.25″ aluminum plug then it could be mounted into the focuser of the telescope rather than the body of the spectroscope, sounds feasible but gives problems with maintaining the critical distance to the collimator lens. A better method is to use a 12 mm/16 mm plywood partition fitted close to the end of the housing baseplate centrally drilled to 1.25″ and fitted with an M3 clamping screw to locate the slit plug in position. This allows the slit to be rotated, to align with the grating grooves and moved back and forth slightly to obtain the best collimator focus. If a gap of approximately 70 mm is provided in the housing cover the after assembly it is still possible to view the front surface of the slit plate at an angle. This comes in handy when aligning the solar image.

The extreme front of the housing, beyond the entrance slit, at the end of the 70 mm gap can then be closed by another 70×70 mm partition this one with a 30 mm central hole. Looking now at the imaging side of the spectrograph, if we use the suggested helical focuser then this should be mounted once again on a 70×70 mm partition with a 30 mm central hole about 50 mm inside the focal point of the lens. This then gives sufficient space for the camera and nosepiece.

The final shape of the base board is now complete. It should be 70 mm wide and extend from the front partition at the entrance slit to the grating, and sit at $\pm 35°$ angle towards the final focuser partition. The 76 mm (70 mm plus the material thickness) sides of the box section can be glued along the edges, the partitions with the components glued in place and the interior carefully painted matte black. The lid section with the 70 mm cut-out immediately before the slit position can be traced around the housing, to give an overlap at the top of the walls to provide sealing and fitted with screws to allow it to be removable for access. The classical spectrograph previously illustrated in Fig. 6.1 has a similar construction.

Independently of the telescope, the spectrograph can now be tested by fitting the camera and pointing the entrance slit towards a bright sky, or bench tested using a fluoro energy saving lamp. By adjusting the slit gap as narrow as possible until it is just a visible thin line, the slit position until it is exactly at the focus of the collimator lens and aligned with the grating grooves, the grating rotation to bring the target region of the spectrum into view, and the camera focus to bring the absorption/emission lines into tight focus, we can verify the assembly is correct and useable.

The last challenge is to fit the spectrograph to the telescope. This demands that the alignment of the telescope optical axis aligns with the center of the entrance slit and the optical axis of the collimator. It also needs to be able to allow the accurate focusing of the solar image onto the entrance slit and hold this position securely. There is a secondary requirement to ensure that the telescope/spectrograph can be safely and securely fitted and balanced onto the mount.

A basic option is to have the telescope mounted in the usual rings on an extended dovetail projecting beyond the rear of the focuser, the spectrograph is then securely fixed to the dovetail and aligned with the telescope optical axis with spacers as necessary. The gap between the telescope focuser and the front partition of the spectrograph, when the solar image is focused onto the entrance slit should be closed using standard extension tubes (either 2″ or 1.25″) or spacers.

The spectrograph locked in place immediately behind the telescope then allows the end of the telescope focuser tube to push against the face of the spectrograph housing's outer partition. By slightly loosening the telescope clamp rings, the body of the telescope will slide within the rings as the telescope focus is changed. When the best focus is achieved the clamping rings can be re-tightened. Section 9.13 shows a slightly more elegant method utilizing a shelf slide and adjusting screw.

This basic classical spectroscope based SHG has the potential to be upgraded to other telescopes as future needs dictate and provides a relatively easy introduction to the world of the SHG.

7.2   Option 2: A Long Focal Length Littrow Design

This design has the capability of providing high dispersion and higher resolution. This then gives significant opportunities to carry out the science aspects of solar imaging, incorporating the Zeeman Effect, magnetometry and more. If full disk imaging was required on smaller cameras (DMK21) it would be necessary to either reduce the telescope focal length or prepare mosaic images, but this would compromise the higher resolution.

The general layout suggested would be a long focus telescope on which a long focus Littrow spectroscope is reverse mounted. This arrangement has the potential to provide a compact layout capable of being supported by medium sized EQ mounts. It would additionally be more suited to the use of larger frame sized cameras.

The Littrow spectrograph design does not easily utilize a magnification reduction, hence the solar image produced by the telescope should match the camera chip size for full disk imaging. In the case of the DMK51 this would restrict the telescope to a focal length of <700 mm. Mete's VHIRSS design in Sect. 9.13 shows an in-line Littrow version based on similar dimensions.

The current concept is an 80 mm aperture telescope focal length of 600 mm (f7.5) combined with a identical telescope as the Littrow spectrograph. The telescopes selected are readily available to the amateur and would provide a reasonable solar image of 600/109 = 5.5 mm diameter. To save money, a cheaper 60 mm, 700 mm focal length (f11.7) beginner telescope could be used for the Littrow. Surplus Shed options might also reduce the cost.

Having pre-selected the optics, we need to consider the grating and the entrance slit sizes. Obviously the slit length needs to be at least 6 mm to match the camera chip size. As we saw in the first example the commercial slits are unsuitable; we definitely need a DIY slit assembly. Based on the DMK51 camera option then the target slit gap would be 10–12 μm.

The f7.5 telescope system combined with the Littrow 600 mm collimator focal length gives an optimum grating size of 80 mm size (600/7.5). Gratings of this size are not readily available to the amateur at reasonable process, so the compromise solution is to use 50 mm square grating size. The 80 mm aperture could easily be stopped down to 50 mm, giving an f12 beam. This will reduce any unwanted reflections and improve contrast. The lost of throughput efficiency, almost 38 % (50/80) is significant but is acceptable for solar imaging. At least an 1800 L/mm grating should be used.

We then have:

Telescope — 80 mm, 600 mm focal length (f7.5)
Slit — 10–12 µm, >12 mm long (DIY)
Littrow — 80 mm 600 mm focal length, stopped down to 50 mm (f12)
Grating — 50 × 50 mm, blazed 5500 Å, 1800 l/mm reflection grating

This configuration, according to the SimSpec SHG spreadsheet, would give a dispersion of 0.03 Å/pixel, a resolution (at Hα −6563 Å) of 0.19 Å (R = 34000) and a wavelength coverage of 53 Å. The anticipated resolution of 0.19 Å is significantly better than that normally achieved with double stacked etalon filters and the control of the CWL selection adds a new dimension to the scientific work which can be carried out.

Just for comparison, if the telescope was increased to a 100 mm aperture telescope focal length of 1200 mm (f12) the calculated results would give a solar image size of 11 mm, with similar dispersion and resolution. The 50 mm grating size would then be close to optimum (600/12), throughput efficiency improved and the solar surface spatial resolution also would be improved. Full disk imaging would require at least the combining of two images into a mosaic.

7.2.1   Construction Notes

The two optical tubes can be located in standard rings fitted to dovetails and mounted in a side by side adaptor plate fitted to the mounting. Alternatively both telescopes can be supported in rings fitted to a strong baseboard (18 mm plywood × 250 mm wide and 650 mm long). A dovetail fitted to the lower surface then allows fitment to the mount.

The simplest method of mounting the grating is probably that used by Mete, illustrated in Sect. 9.13. The grating sitting in a plastic holder clamped central on a M5 threaded screwed rod across the dewcap enclosure of the Littrow telescope. The external tangent arm providing the grating positional control.

A significant factor in this design is that both the telescope and the Littrow telescope will have identical chromatic aberrations. This then allows the entrance slit plate to be mounted on a carriage behind and between the telescope focusers using two front mirrors to align the optical axis. As the slit assembly is moved to re-focus the telescope, it automatically re-focuses the Littrow by the same amount.

This is very similar to the system employed by Rousselle as detailed in Chap. 9. The two small front surface mirrors can be 20 mm elliptical secondary mirrors (readily available from your local astronomy supplier) or just 25 mm square front surface mirrors mounted at 45° in a 6 mm plywood frame with centre distance to match the two telescopes.

In the middle of this frame a partition holds the slit assembly allowing for slight radial adjustment. See the Rondi/Rousselle/Poupeau solutions in Chap. 9. The support frame then sits on a slide mechanism which allows at least a 5 mm travel to provide focus adjustment between the blue/green/red regions of the spectrum. Ideally this would be moved using a calibrated micrometer, or at least a fine threaded screw. The base of the slide would then sit on the support board between the telescopes.

It may seem that it would be easier to just to use a couple of star diagonals with the slit plate sandwiched between them. The diagonal assembly could then be mounted into the telescope/Littrow focuser(s) and the focusing mechanism of the telescope used to focus the solar image on the entrance slit. The reality, however, is that the mirrors mounted in the typical star diagonal may not be sitting at exactly 45° and the amount of re-adjustment thus required becomes a nightmare. Sure, it can be done, but needs a lot of patience and is better avoided in the first place.

The final assembly of telescope/Slit plate/Littrow can be checked for optical alignment using a laser collimator temporarily mounted at the centre of, and square to, the telescope objective. This Littrow design represents an ideal, compact SHG arrangement and gives a performance on solar disk coverage and resolution which is close to optimum.

7.3   Final Assembly Checks for the SHG

The notes in this section are directly applicable to a digital SHG constructed with the classical spectrograph but can be equally applied or adapted to any other chosen design.

Assemble and test the spectrograph section of the SHG first. Ensure that the entrance slit is securely located and well focused by the collimator. The length of the slit gap must be parallel with the grooves on the grating. The distance from the collimator lens to the grating should be kept to a minimum. The grating, accurately centered on the optical axis, with the front surface aligned with the axis of rotation should be able to be rotated smoothly and to be locked into a final position to suit the target wavelength. The imaging lens, focused on infinity, should show the solar absorption lines in clear, tight contrast. Again, the imaging lens should be positioned as close to the grating as physically possible. The imaging camera and its focusing mechanism must be securely located with no movement or slop.

The horizontal or vertical axis of the CCD chip should be aligned with the axis of the spectral image and show a clean tight well focused image of the absorption lines and the edge(s) of the solar disk, generally as seen at the top and/or bottom edge of the spectral image.

The spectrograph is then fitted to the telescope. The attachment must be solid with no slack or sag. Using the telescope focuser only the solar image must be able to be brought to a tight focus on the front of the entrance slit and the edge(s) of the solar disk also appear clean and in good focus.

Ideally when using an EQ mount it will be polar aligned (PA). This can be done by aligning the polar axis N-S close to the time of solar meridian passage (check your local astronomical handbook for local timing) or by re-using marked positions on the ground from more rigorous night-time polar alignment. The closer the PA the less the Dec drift will be during the scan and exposure.

The complete SHG should sit well balanced on the mount with the slit length aligned with the Dec axis. Check the RA/Dec movements to make sure they allow the solar disk to be drifted onto and across the slit gap.

A final comment: The performance of the SHG will be significantly improved when the housing is sealed and all the optics are clean and effectively baffled. Look for unwanted parasitic light or reflections and remove them.

7.4   Calibration and Spectral Resolution

Measuring absorption lines in the solar spectrum can lead to an incorrect assessment of the resolution. Some of the solar lines are very broad—the Hα core is over 2 Å wide—and it's much better to use a neon reference lamp to generate very narrow emission lines. The typical neon lamp spectrum is shown below in Fig. 7.1.

The reference spectrum can then be used to calibrate the dispersion of your spectrum in wavelength per pixel (Å/pixel) and also allow the determination of the actual resolution (Å) of the spectrograph. Any suitable neon indicator bulb can be used as a reference lamp.

7.4.1   Method

Assemble and set-up the SHG, and point the telescope towards a bright daytime sky. Rotate the grating to bring the Hα line near the centre of the CCD frame. Once that has been achieved do not move the grating position. Reposition the SHG to point the telescope towards the neon lamp, which should be positioned on the optical axis as close to the objective as possible. Taking a short AVI file should then show the neon emission lines around the Hα wavelength. Note which side of the image is towards the blue end of the spectrum. As the grating is rotated it will move the spectral image towards the red end of the spectrum. Convert the AVI to a series of .bmp files and select one of images as a master reference image. This should

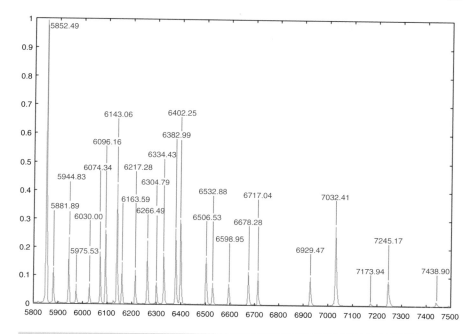

Fig. 7.1 Neon reference lamp spectrum (Buil)

show clearly the series of well-focused neon lines without over exposure. The image should be orientated with the blue side of the spectrum towards the left hand side. Flip the image if necessary.

Using freeware software like BASS Project or VSpec (discussed in Chap. 8) allows us to now to import and calibrate the master image. See the example illustrated in Fig. 7.2

Based on calibrating the master image using the BASS Project software (Sect. 8.11.1):

1. Use the tab *File/Add/Open images* and select the master image. This will display the master image at the top of the screen (marked "01") and the associated profile below.
2. Use the tabs Chart Settings/Advanced to change the wavelength units to Å, then the tab *X Axis/X Axis calibrated text* change to Å.
3. Look at the reference illustration Fig. 7.1 and attempt to identify the sequence of the neon lines around the Hα wavelength (6563 Å)—notice there is a pattern of two prominent neon lines then one line then a further two prominent neon lines straddling the Hα position (6506.53/6532.88/6598.95/6678.28/67 17.04 Å).

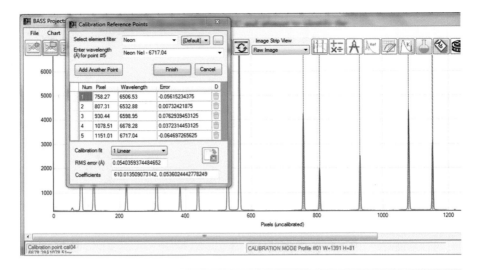

Fig. 7.2 Neon calibration (BASS Project)

4. Use the tab *Calibration/New Calibration*. Click and drag the mouse to select the left hand line. In the drop down screen, *Select element*—select neon, then under *Enter wavelength* select the wavelength for the selected line. Press *Add another point* and repeat for as many lines as possible.
5. The information bar at the bottom of the screen will display the dispersion in A/pixel. Note this result, it will assist you when selecting the CWL for the spectroheliogram later.
6. To obtain the FWHM result, use the tab Tools/Measurements and elements. Then on the drop down screen the Measurement options tick the FWHM box and using the mouse click and drag to select a neon line close to the Hα position. Finally press *Run* and the screen will display all the information on the selected line including the FWHM and the resolution i.e. FWHM 2.17 Å (R = 3040 at 6599 Å).

The dispersion and the FWHM results give a very good first approximation of the number of pixels which should be selected in the processing of the spectroheliogram to obtain the resolution of your SHG. A dispersion of 0.7 Å/pixel and FWHM 2.17 Å would indicate a selection strip width of 3 pixels, giving just over 2 Å bandwidth. Choosing less will cause under-sampling, selecting more will increase the bandwidth. Refer also to Fig. 6.6 which shows the typical actual FWHM resolution measured can be 2–4 times the dispersion.

7.5   Taking the AVI Video

Obtaining the best AVI files possible for subsequent processing requires a bit of pre-planning. The SHG should be tested, rigidly mounted, balanced and capable of aligning to the Sun. Calibration of the grating position in wavelength should have been established and the settings required to locate the wavelengths of interest noted. The entrance slit gap needs to be pre-set to suit your telescope spectrograph and camera.

With the imaging camera attached and connected to your computer, point the telescope towards the bright sky and obtain an image of the solar spectrum. Setting the AVI codec to Y800 is a safe option. Check that the solar absorption lines are visible: the top and bottom edge of the spectrum may at this stage appear fuzzy. Refocus the camera only (this may involve also changing the gain setting of the camera) to achieve crisp and clear lines. Finally, re-set the grating to the target wavelength and re-check the focus. This may take a few iterations to get perfect. The better the spectral focus, the better the outcome.

While initially imaging the target line, also confirm the orientation of the spectrum both in wavelength and against the horizontal axis of the CCD tilt. The red or blue sides of the spectral image can be checked and verified using nearby absorption lines. Use one of the atlases suggested in Sect. 5.4 to determine which direction in your image, left or right, is towards the blue side of the spectrum. It is the convention in spectroscopy that the blue, shorter wavelengths are positioned towards the left hand side. Correcting spectral orientation and any tilt can be done at the same time by simply rotating the camera relative to the spectrograph. Don't worry about the lines appearing to slant within the spectrum; this is normal and can be corrected during subsequent processing.

Before moving on, experiment with various camera settings, exposure, frame rate, gain, etc. This is best done using the Hα line. What we want to achieve is a well exposed line to record the depth of detail in the line. This may mean that the continuum at either side looks overexposed and white. Figure. 7.3 shows a typical good exposure setting.

Fig. 7.3 Well exposed absorption Hα line

The next task is to position the solar disk in focus on the entrance slit. Carefully point the telescope towards the Sun; when the solar disk enters the slit gap it will be immediately obvious due to the increased brightness. As the edge comes into view the bright spectrum will be a very narrow line. This spectral line will be seen to increase in height as the solar image is moved across the slit gap before starting to reduce in size as we leave the opposite edge of the Sun.

Look carefully at the top/bottom edge of the solar spectrum. These edges will represent the edge of the solar disk image. Use the mount's drive to reposition the spectral band if necessary. Carefully, start focusing the telescope, and do not touch the spectrograph focusing until the edges appear as sharp as possible. Repeat this cycle a couple of time to ensure best focus is being achieved.

The last step is to verify the scanning position of the solar disk relative to the entrance slit gap. Position the solar spectrum across the center of the camera and use the mount slow motions in RA to move the solar image back and forth across the slit. Depending on your arrangement the whole of the solar disk may not be visible in your slit. If that is the case you will have to do two or more scans to gather the image data and later, during processing, combine the images to make a full disk image.

Move the solar image in the slit gap using the Dec slow motions to select the zones to be scanned e.g. you may have a bottom scan, a middle scan and a top scan. Finally, double check that both the absorption lines and the edges of the spectrum are crisp and sharp. Briefly switch off the mount RA drive and note the direction of the drift.

Now it's almost time to gather the scanned images. Reposition the solar image, using the mount controls to position it just off the slit gap to the East so it will drift across the slit gap when the drive is switched off. This may sound very complex, but with some practice becomes second nature, and only takes a minute or two to carry out. Stop the mount RA drive and start the recording with the camera. Make sure you are set to the exposure/frame rate/gain readings you previously noted.

As the Sun drifts onto the slit gap a narrow spectral band will appear. This gradually grows in height as the exposure continues until we reach the center of the solar disk. At this point the height of the spectral image will be at maximum. As the exposure continues the height of the spectral image will start to decrease until finally the image of the Sun moves off the slit gap.

The full scan across the solar disk will take 120 s. Depending on your camera frame rate you may collect up to 2000 frames in your AVI. The size of the AVI file will be large, possibly 1–2 Gb.

In the next chapter we will look at the range of software available to allow us to process the AVI into a final spectroheliogram image. That's when things start to get exciting!

Further Reading

Harrison, K.M.: Astronomical Spectroscopy for Amateurs. Springer (2011)

Webpages

http://www.sciencecenter.net/hutech/borg/focuser.htm
http://www.optcorp.com/telescope-accessories/focusing/ba-eyehold-2-1-25inch-t2-adapter-w-microfocuser.html

Chapter 8

SHG Processing Software

Having obtained an AVI file centered on the target wavelength it is now necessary to extract and then recombine the image strips selected at the required wavelength and selected bandwidth from the AVI file. There are various freeware software options available to assist us in this task.

Each spectrograph arrangement will produce different aberrations and the AVI file is likely to show effects like tilt, slant, and smile. These need to be corrected before a successful line strip can be extracted and used to produce a spectroheliogram.

8.1 Tilt

When the horizontal or vertical axis of the CCD chip doesn't align exactly with the axis of the spectrum being produced it is said to be tilted (Fig. 8.1). This can usually be easily corrected by rotating the camera body relative to the spectrograph. It is highly recommended that the tilt angle be reduced to a minimum before imaging.

8.2 Slant

When optical layouts like the Littrow or Ebert–Fastie are used, the absorption lines recorded will be seen to be slanted within the spectral band as shown in Fig. 8.2.

© Springer International Publishing Switzerland 2016
K.M. Harrison, *Imaging Sunlight Using a Digital Spectroheliograph*,
The Patrick Moore Practical Astronomy Series, DOI 10.1007/978-3-319-24874-5_8

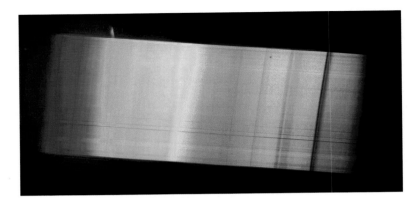

Fig. 8.1 Spectrum shown tilted across CCD frame

8.3   Smile

The off axis effect on the extreme edges of a large solar image on a long slit can cause the absorption lines to appear curved in the spectrum. You can see the minor smile effect starting to show in Fig. 8.2. Any software used to prepare the SHG mosaic images should be capable of correcting or compensating for these aberrations.

Obtaining a SHG Mosaic Image (Spectroheliogram): The Conversion Process Step by Step

- Crop the AVI file to an area surrounding the widest part of your spectral image, showing the full height of the solar disk being recorded including some background to catch any prominences at the target wavelength. Allow 20 pixels or so at either side of your target absorption line. This will dramatically reduce the processing time.
- Apply corrections (tilt/slant/smile) as necessary to present horizontal (or vertical, depending on the scan direction) absorption bands.
- Select the bandwidth required in pixel width and the wavelength center (red/blue wing etc.).
- Extract this strip to a series of images. Some programs do this automatically and prepare the final mosaic.
- Combine the strips by laying them up side by side into a mosaic covering the surface image.
- Re-scale the raw image to correct aspect ratio and make the solar image circular.
- Feather the solar edge to remove the slit steps.
- Apply synthetic flats to improve the cosmetics; this may mean the removal of transversalium lines, or even upping the surface brightness and darkening the background).

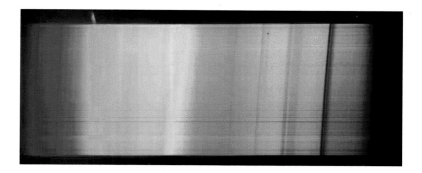

Fig. 8.2 Image showing slant of absorption lines within spectrum

- Apply some histogram contrast adjustments.
- Apply some small sharpening—to enhance detail.
- Colorize the image as required.

> **Note:** It helps if you have pre-calibrated your spectral images in wavelength (see Sect. 7.4 above). Knowing the wavelength by pixel number and dispersion, will help you to locate exactly the central wavelength and determine pixel widths for best resolution.

8.4   Flats

Similar to conventional astronomical imaging, the use of flats can improve the quality of the spectrum recorded. Flats can be taken using a flat spectrum Halogen lamp. These should be re-done for any subsequent change in the optical configuration or change of target wavelength.

8.4.1   Spectroscopic Flats

Normally, as we've already seen, any light entering the spectrograph will record as a spectrum image. The issue is how to convert this to a usable flat image.

There are two parts to the solution. The first is to use a light source which generates a continuum spectrum with no emission or absorption lines, and the second requirement is to ensure that there is no movement of the grating and that the flat is taken at exactly the same setting as the spectral image.

Fig. 8.3 Taking spectral flats

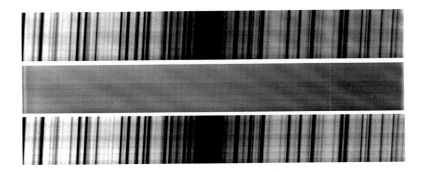

Fig. 8.4 Spectral flat Corrections (Buil)

A suitable light to use is a Quartz halogen lamp (20 W), white high intensity LEDs or an Electroluminescent Panel. These provide a very uniform spectrum. Directed to a sheet of white paper or white T shirt draped over the telescope objective, this light will give a usable flat (see Fig. 8.3).

Figure 8.4 shows the application of spectral flats. The top image is the raw spectrum as recorded. The middle image is a flat image obtained at the same grating settings and the lower image the final corrected spectrum.

Remember, a flat taken in the red region will not be usable as a correction flat image in the blue region. Any change to the spectroscope configuration, whether change of telescope focal ratio, grating, camera distance, camera lens, or CCD, will require new correction images. The flat corrections are applied by dividing the spectral image by the flat image.

The above process can suppress the vignetting, doughnuts and pixel response in a conventional spectral image. Where the SHG spectral data is recorded frame by frame in an AVI file, applying the flat field correction is more problematic. Identical regions from the CCD images (X-Y dimensions and X central wavelength) must be used to prepare the final image mosaic prior to applying the resulting flat frame. Due to the difficulty of achieving the above, most SHG users just rely on applying synthetic flats to correct the most obvious defects.

8.4.2 Synthetic Flats

Synthetic flats can be applied to the final mosaic spectroheliogram image to reduce the transversalium lines in the mosaic image caused by slit edge defects or dust. Good synthetic flats can be prepared by analyzing the spectral image and making a correction profile or by using the imaging software to extract the image detail and only leaving the transversalium features. Your imaging software (e.g. AstroArt V5) may also have the ability of X Axis Binning. This function can be used to prepare a useful synthetic flat image when the transversalium is orientated horizontally across the image. Duplicate the image, apply X Binning and then divide the original by the result (see Fig. 8.5).

The use and application of synthetic flats will be discussed further when reviewing the individual processing software options.

Fig. 8.5 Using an X binned synthetic flat (Mete)

8.5   Processing Software for Digital Spectroheliograms

The introduction of the digital camera to the SHG brought with it the new challenge of being able to process a recorded spectral AVI into a final spectroheliogram. At the time existing digital imaging software was still in its infancy and certainly there were none which could handle the task. Defourneau at the same time as pioneering the digital SHG developed the first software capable of doing the job, his SpecHelio package. Buil quickly followed with SHG enhancements in the IRIS astronomical imaging software.

Later the highly capable Virtual Dub and ImageJ software were introduced and the past two years have seen further SHG software developed by talented amateurs like Wah-Heung Yuen (Spectral Line Merge) and John Paraskeva (BASS Project) who are committed to the ongoing support of the digital SHG.

You will see by comparing the results obtained from each of the software packages that they are well capable of producing some excellent results. All of the current digital SHG processing software is distributed as freeware and is MS Windows based. They have all been trialed using a computer running Win7, 64bit.

8.5.1   SpecHelio Bas

This program developed by Defourneau takes a sequence of BMP files and reconstructs a mosaic image (Fig. 8.6). AVI files have to be converted to a series of BMP files before processing. There are many freeware programs available on the internet to do this conversion, such as VDub for one.

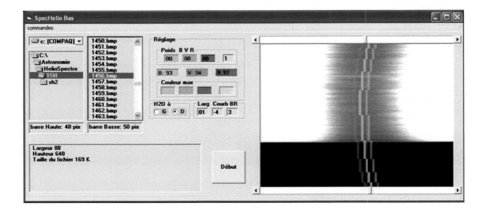

Fig. 8.6 SpecHelio Bas—main screen

The freeware Avi2Bmp conversion software is included in the program down-load package. When the AVI file is converted SpecHelio Bas will handle up to 4000 frames. Open the SpecHelio Bas program and enter the file location in the right hand window. Check that the files to be converted are in sequential order within the folder, i.e. file0001, file0002 etc.

Select one of the listed files, making sure it has the maximum spectrum height. The file name will then show in the bottom window. Click in the large image win-dow and the selected frame will appear. Move the cursors at the top and bottom of the image frame to center the line which interests you. See Fig. 8.7 for an example of the display at this point.

A white line is registered on the spectrum by moving the cursor and clicking the mouse will place this line where desired on the heart of the line. Chances are that your spectrum will have a curve (smile), this is normal.

In the box named "*Courb*" in "*Larg, Courb, Br*", register a value in pixel, for example—4. Click again, the white line curves. This represents the line of the extraction from the spectrum. A negative number forms a convex curve, a positive a concave curve.

Registering the number 6 in the box "*Larg*" will result in the program making 6 sequential cuts and will take the average of them. Taking a series of pixels to broaden the cut will significantly reduce the noise in the image.

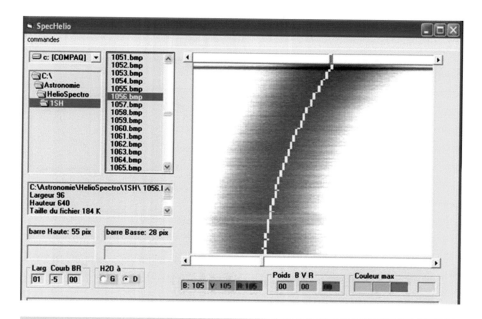

Fig. 8.7 Working screen (SpecHelio)

If we register a pixel value in box "*Br*", we enter the Doppler mode. After a click on the image, the white line is now divided into 3 lines: Blue (B), Yellow (V), and Red (R). Each is spaced at the nominated pixel value entered. It is in these three columns that the cuts will be made. These three operations can be combined.

In lower part of the screen, there are two colored check boxes, green and red. The box to be marked depends on the orientation of your camera. It is used to choose the blue or red code to represent side at which the prominent H_2O line lies relative to the Hα line. When your choice is made, click on the *Debut* button. Calculation starts and the image will be constructed—this may take a few seconds.

Note that the completed, reconstituted spectroheliogram image file (Fig. 8.8) is automatically named "SH.BMP" and is saved in the same folder as the main program. This is shown in the window for reference.

The boxes color B, V, R indicate during calculation the maximum value for the colors Those are coded for the BMP from 0 to 255 in each box "*Poids BVR*" is used for to make mini corrections on each color to improve the color balance.

The gray box at the right hand side is a multiplier factor which affects all the colors. If you register a multiplying factor of 4 or 5 the solar disk will be saturated and becomes black. On the other hand the prominences, otherwise invisible, will appear around a black disk—a made-to-order coronograph.

Results

Figure 8.8 shows the final spectroheliogram mosaic obtained.

Fig. 8.8 Result (SpecHelio)

8.5.2 IRIS

IRIS is one of Christian Buil's older image processing programs and has been around for many years (see Fig. 8.9). It can handle webcam (640×480 pixel) AVI files up to 2 Gb.

One feature of interest is the ability to convert webcam AVI files to a series of Fits images. The command "Scan2pic" allows the selection of a nominated X column to be extracted (Fig. 8.10) from the first frame and appends the same column from each successive frame to form a mosaic.

There is also another very useful command, "Win_Webcam," which can be used to nominate a region of interest, for instance an area a few columns wide at the selected wavelength. This significantly reduces the size of the files being manipulated.

The Scan2pic command doesn't seem to like long file names, for instance C:\myfiles\SHG\images\IRIS\Scan2Pic. Locating the IRIS software in the root directory seems to work, i.e. C:\IRIS. Also, since the frames are extracted as .fit files they can't be used by other reconstruction programs without first converting them to .bmp files. The saved files must be named. i.e. Solar1, Solar2 etc.

Unfortunately IRIS only extracts a single pixel column for the image reconstruction (Fig. 8.11). Using a single pixel column gives significant under-sampling.

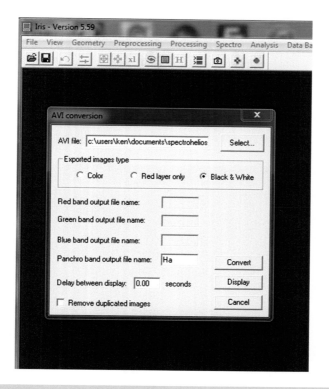

Fig. 8.9 IRIS working screen

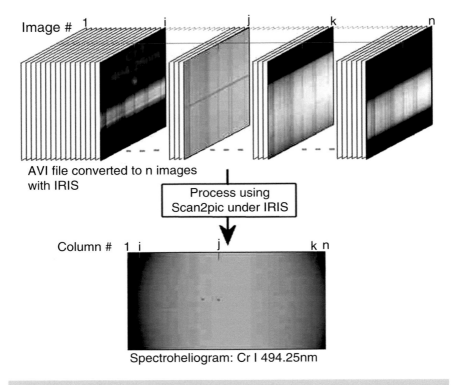

Image # 1   i   j   k   n

AVI file converted to n images
with IRIS

Process using
Scan2pic under IRIS

Column #   1   i   j   k   n

Spectroheliogram: Cr I 494.25nm

Fig. 8.10 Applying Scan2Pic (after Rousselle)

Results

Fig. 8.11 shows the final image produced by IRIS.

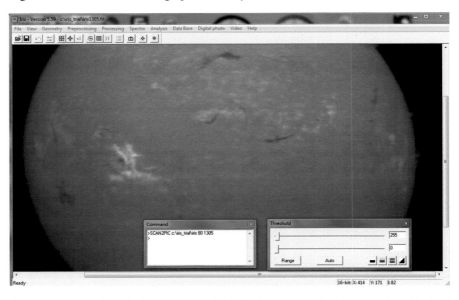

Fig. 8.11 Final image mosaic (IRIS)

8.5.3   Virtual Dub/ImageJ

Many digital SHG users have used a combination of Virtual Dub (VDub) and ImageJ software to process their images. The AVI file from the camera is opened in VDub (Fig. 8.12) and the region around the target wavelength cropped to produce a much smaller, more manageable set of files. Using the *rotation* option both tilt and slant can be corrected.

The selected area based on the target CWL and a pixel width is then exported as a series of BMP files to your nominated folder. There doesn't seem to be a way of correcting for smile, a limitation of the program. See Fig. 8.13.

The second stage of the process is to load the BMP files into ImageJ and select the column position and width to be assembled into a spectroheliogram mosaic (see Figs. 8.14, 8.15 and 8.16).

The resulting mosaic can be further processed to improve the aspect ratio. The end result is your digital SHG image of the Sun as shown in Fig. 8.17.

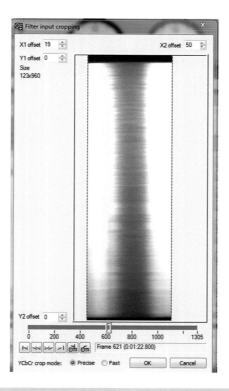

Fig. 8.12 VDub selection screen

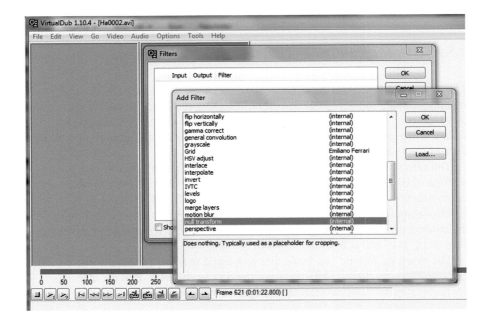

Fig. 8.13 VDub AVI processing screen

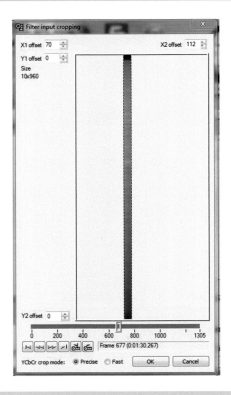

Fig. 8.14 VDub selection of crop zone (example shows 10 pixel wide selection strip)

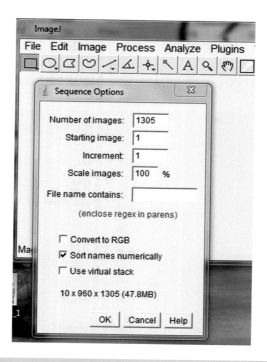

Fig. 8.15 ImageJ processing screen

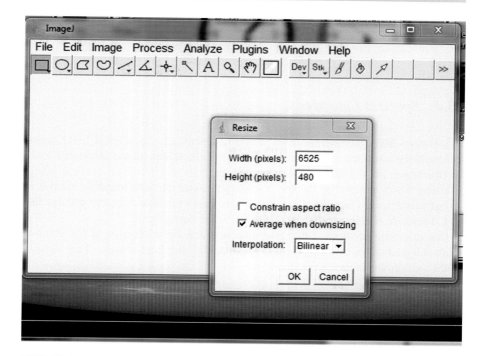

Fig. 8.16 ImageJ resizing screen to correct the aspect ratio

Results

As shown in Fig. 8.17 the result obtained is very clear and the contrast is maintained.

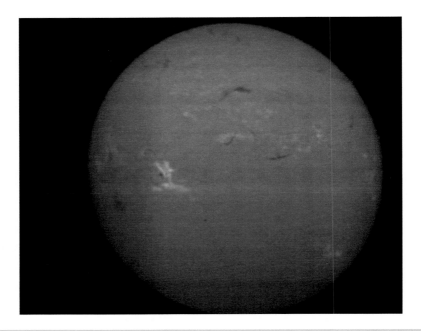

Fig. 8.17 Final reconstituted image (ImageJ)

8.5.4   Spectral Line Merge

Wah-Heung Yuen has written this program specifically for digital SHG users. Although the interface is very basic (Fig. 8.18) it contains all the necessary tools for conversion of AVI files to a finished mosaic.

Using VDub to pre-crop the spectral image around the wavelength of interest is almost mandatory. This step dramatically decreases the time taken to process the AVI. There are options available for tilt and smile corrections.

One interesting feature is that Wah has included the conversion of the column width selected down to a single pixel in producing the final mosaic. This significantly reduces the aspect ratio effects.

Synthetic Flats

Wah makes use of Photoshop to prepare and apply synthetic flats. This is well explained in his tutorial and the reader is referred to these documents for more detail. In summary, he flattens the solar disk image by multiple copying of the

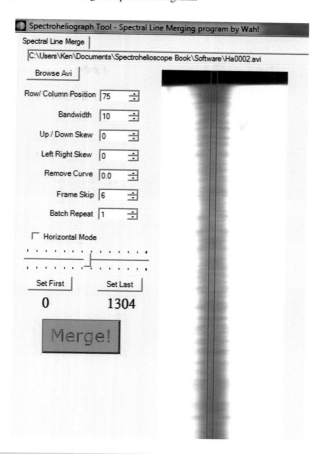

Fig. 8.18 Spectral Line Merge screen

image and changing the luminosity each time (darkening for CaK and lightening for Hα) until most of the surface detail is suppressed.

An area with maximum height and minimal surface detail is then selected and transformed to 1 pixel wide (see Fig. 8.19). This is then further transformed to full picture width and the histogram adjusted to 0–255 range (Fig. 8.20). The original image is then divided by the final synthetic flat to remove the transversalium. This results in a similar synthetic flat to that obtained by the X binning method.

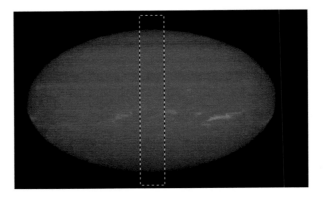

Fig. 8.19 Selection of the synthetic flattening zone (Spectral Line Merge)

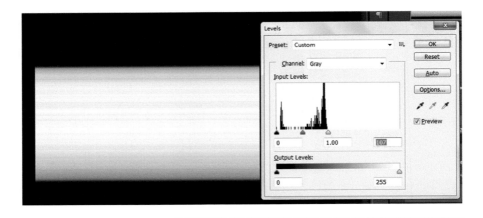

Fig. 8.20 Final synthetic flat (Spectral Line Merge)

Besides transversalium removal, Wah also uses similar filter techniques to improve the dark sky background, and remove any obvious gradients within the image.

Results

The use of this synthetic flat method effectively suppresses the transversalium and the resulting image shows the Hα detail very well (see Fig. 8.21).

Fig. 8.21 Final processed SHG image (Spectral Line Merge)

8.5.5   BASS Project

BASS Project is a freeware software package (Fig. 8.22) developed by John Paraskeva for amateur astronomical spectroscopy. It was upgraded in 2015 to V1.8 to include some very capable spectroheliogram processing features.

Any AVI file can be imported to BASS Project, and it can handle corrections for tilt, slant and smile. The final image can be filtered using a synthetic flat to remove transversalium lines. After loading BASS project, open the *TOOLS/RECONSTRUCT FROM SPECTRA* tab.

Load in the AVI video to be used (see Fig. 8.23).

Using the slider, find a frame near the middle of the sequence with transversalium lines and use this to measure tilt angle.

Similarly, use the slider to find a frame in the middle with a prominent absorption line next to your target line and use this to measure slant angle (see Fig. 8.24). Enter these tilt and slant angles in the boxes and press *APPLY*. Tick the Crop option and draw area to be cropped (on the max height frame).

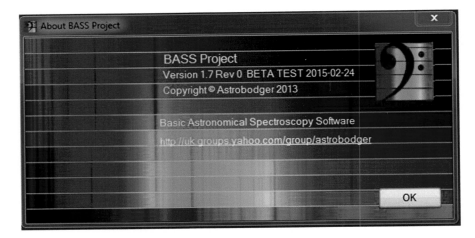

Fig. 8.22 BASS Project

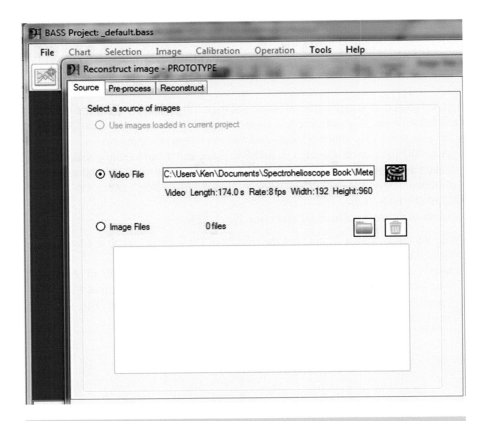

Fig. 8.23 AVI Load (BASS Project)

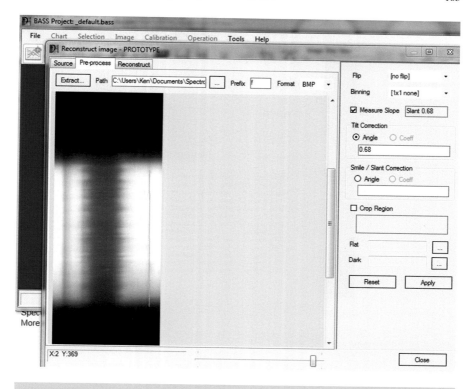

Fig. 8.24 Measuring the slant angle (BASS Project)

Click *EXTRACT* button, specify extract path etc and extract the cropped frames
   Fig. 8.25. Reset the settings so they are blank (to avoid applying correction
   again). Then load the extracted image files.
Move the mouse to the middle of the Hα line; note the X value at midpoint.
Click the *RECONSTRUCT* tab.
By entering a range of values on either side of the mid x pixel value, we can change
   the effective bandwidth being recorded. The example in Fig. 8.26 has the mid-line
   at pixel 30, and a bandwidth based on 10 pixels. The result is shown in Fig. 8.27.

Synthetic Flats

Synthetic flats can also be generated within the BASS project software. These can
then be used to remove the slit defects—transversalium.

   The *PREPROCESS* tab is used to view the frames and locate one which shows
transversalium, also close to the maximum spectral height. This is normally found
near the middle of the sequence.

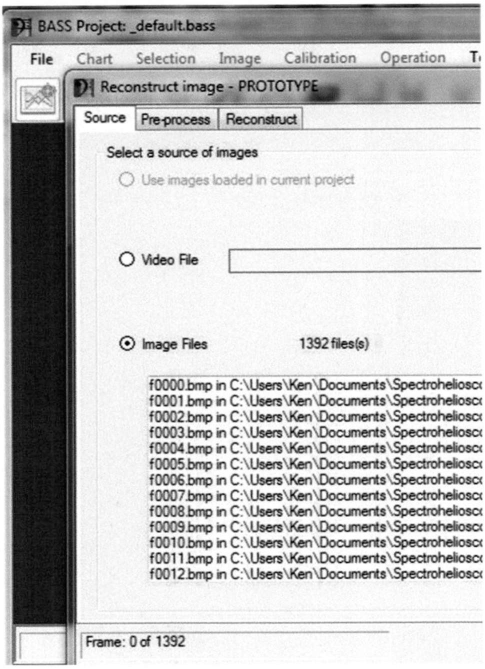

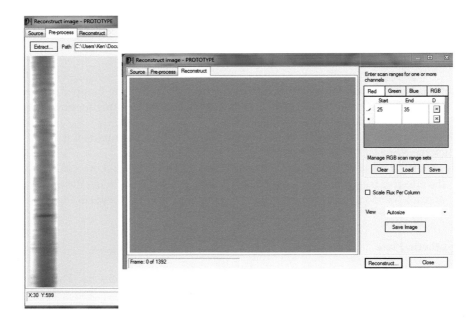

Fig. 8.25 File sequence (BASS Project)

Fig. 8.26 Center of Hα line at pixel=30 and bandwidth of 10 pixel selected (pixels 25–35) (BASS Project)

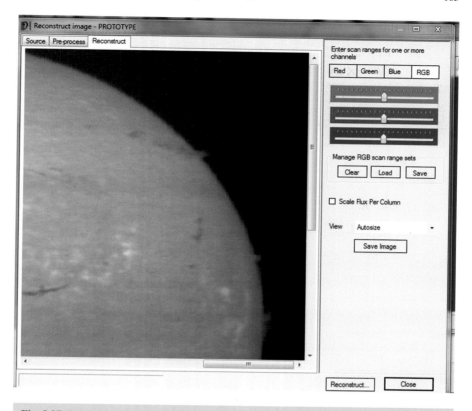

Fig. 8.27 Reconstructed spectroheliogram image (BASS Project)

Open the selected frame into BASS Project (*FILE/ADD/OPEN IMAGES*) and rotate it clockwise (*IMAGE/FLIP/FLIP RIGHT (90)*).

Select an active area outside your target line of interest (*SELECTION/SELECT ACTIVE BINNING REGION*) which includes a section of the image with prominent transversalium lines, seen as vertical dark lines in Fig. 8.28.

Save the resulting spectrum as a 1D fits using the 1D Icon in the header.

The synthetic flat can then be applied to the selected frames using the same processing as above. To do so add this flat file to the "Flat" box in the pre-processing screen, bottom left hand side in Fig. 8.29 prior to the final extraction and reconstruction of the image.

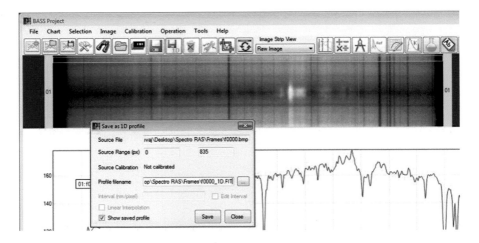

Fig. 8.28 Example of a spectrum with pronounced transversalium—*vertical dark lines* in the image. (BASS Project)

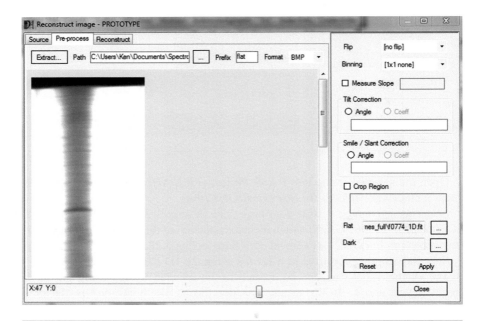

Fig. 8.29 Applying the flat (BASS Project)

Results

Figure 8.30 shows the benefits of the application of the BASS Project synthetic flats. The mosaic result is very clean and the detail crisp. The method used by BASS Project to generate the synthetic flat appears to give an excellent outcome.

Fig. 8.30 Final corrected image mosaic (BASS Project)

8.6 Other Software

There are other software packages available to assist in the processing of the spectroheliograms, some of the more interesting and useful are highlighted in this section.

8.6.1 AstroSnap

The shareware software AstroSnapV2.2 (Fig. 8.31) has been widely used for astro imaging with webcams. It was developed for WinXP but seems to work under Win7 run as Administrator. It also can provide basic processing for spectroheliograms.

A registered copy of the Shareware version of AstroSnap V2.2 is required to access the processing features for SHG images.

By setting up the dimensions of the acquisition strip (2× camera height pixels is recommended) and positioning the solar image at the entrance slit, the results are almost automatic (see Fig. 8.32).

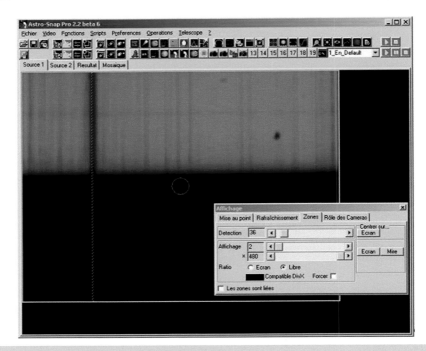

Fig. 8.31 Positioning the acquisition strip on Hα (Astrosnap)

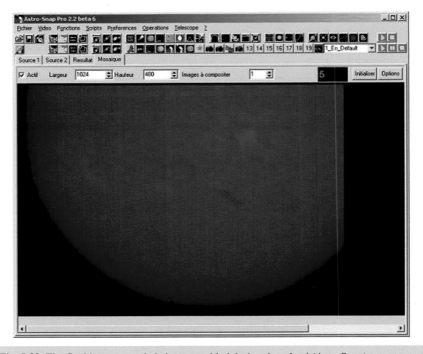

Fig. 8.32 The final image mosaic being assembled during download (AstroSnap)

The program appears to recognize WDM camera drivers for the webcam type cameras (640×480 pixels) and gives almost a "live view" image as the selected strips are assembled into the final mosaic on the screen in real time. The Webpage link provided contains a good English tutorial. Unfortunately it appears that technical support for this program is no longer available. Notwithstanding that it's an interesting program to experiment with.

8.6.2 Spectral Processing Software

With the digital SHG we are recording a section of the solar spectrum based on the target CWL. The resulting spectral image can also be processed and analyzed to determine profiles and wavelengths of the absorption features, including Doppler shift. The BASS project software mentioned above and Valerie Desnoux's VSpec software (Fig. 8.33) provide all the processing capabilities you need to obtain wavelength calibrations of the raw spectrum and prepare a 1D profile for further analysis.

These programs also allow the automatic calculation of dispersion (Å/pixel), the FWHM of any nominated spectral line and the equivalent width (EW) for line comparison.

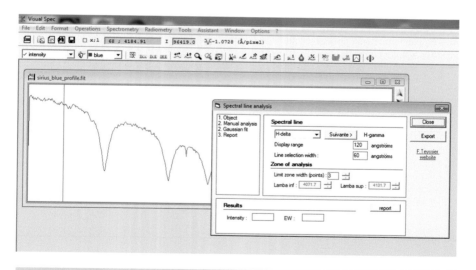

Fig. 8.33 VSpec screen

8.6.3   Mosaic Software

When a small slit length is used on a longer focal length SHG, the height of the solar disk cannot be fully imaged in a single exposure. To obtain a full disk image we need to acquire two or three spectroheliogram images and then combine them to form a complete mosaic image of the Sun. Many amateur astronomers already use Adobe Systems Photoshop CS5/6 for the processing of their astronomical images and this can also be used to produce solar mosaics. The Microsoft Image Composite Editor (MS ICE) is also popular and can achieve good results. Other alternatives include Photostitcher from Teorex and AutoPano from Kolor.

In the next chapter, we will review examples of the spectroheliographs being constructed by various amateurs around the world.

Webpages

http://www.astrosurf.com/buil/us/iris/iris.htm
http://www.virtualdub.org/
http://imagej.nih.gov/ij/
https://uk.groups.yahoo.com/neo/groups/astrobodger/info
http://www.astrosurf.com/cieldelabrie/sphelio.htm
http://www.astrosurf.com/cieldelabrie/sphelio/program/SpecHelio/
http://www.astrosnap.com/
http://www.astrosnap.com/softprotest/Tutorial%20SpectroHelio_Uk.pdf
http://www.astrosurf.com/vdesnoux/
https://www.photostitcher.com/
http://research.microsoft.com/en-us/um/redmond/projects/ice/
http://www.kolor.com/autopano/autopano-features/

Chapter 9

Amateur Digital SHG Instruments

In this chapter we will review the design, construction and performance of some of the best current SHGs being used by the amateur. At the moment there are active SHGs in use in the USA, UK, France, Italy, China and Australia with many more on the drawing board.

The majority of amateur digital SHGs in use today are equatorially mounted. The larger, longer focal length designs use either heavy-duty custom EQ mounts or robust DIY fork mounts. The smaller compact instruments are usually mounted on commercially available equatorial mounts, like the Skywatcher HEQ5 or NEQ6 series.

The exception to this is the Bartolick instrument which is a digital upgrade of his original SHS and fed by a cœlostat.

Andre Rondi has been at the forefront of the digital SHG revolution for many years. He first constructed his SHG in 1998, and later adapted it to record the spectrum with the webcam (see Fig. 9.1).

Figure 9.2 shows the general optical arrangement of the SHG.

The telescope (L) is mounted on a micrometer driven aluminum rail to allow precise focussing of the solar image onto the entrance slit (2). The spectrograph optics are mounted in a square baffled wooden framed enclosure (Fig. 9.3) in an Ebert–Fastie arrangement.

© Springer International Publishing Switzerland 2016   171
K.M. Harrison, *Imaging Sunlight Using a Digital Spectroheliograph*,
The Patrick Moore Practical Astronomy Series, DOI 10.1007/978-3-319-24874-5_9

Fig. 9.1 Digital SHG mounted on a Gemini 40 mount (Rondi)

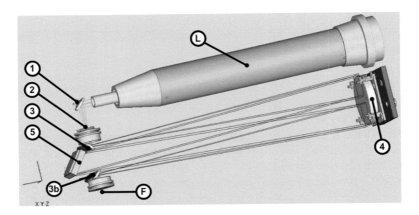

Fig. 9.2 Optical layout of the SHG (Rondi). (*L*) Achromatic refractor, 120 mm, fl 1000 mm (f8.3); (*F*) Mounting plate for camera; (*1*) Front surface mirror; (*2*) Entrance slit assembly; (*3/3b*) Front surface mirrors; (*4*) −155 mm parabolic mirror, focal length 954 mm (f6.3); (*5*) Blazed reflection grating 1800 l/mm

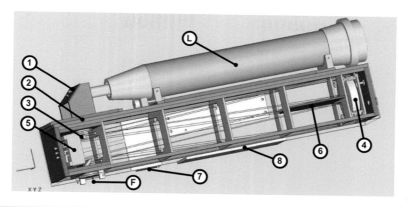

Fig. 9.3 Mechanical arrangement of the SHG (Rondi)

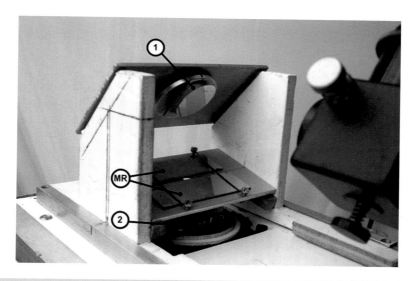

Fig. 9.4 Split front surface mirror acting as a heat reflector in front of the slit assembly (Rondi)

Two sets of mounting plates are provided (8) on the sides of the box construction. This allows the entrance slit gap to be orientated either in RA or along the Dec axis.

The entrance slit is protected by a split reflective mirror as shown in detail in Fig. 9.4. The first surface mirror (1) reflects the solar image down to the split mirror (MR). The gap between the mirrors is set slightly wider than the entrance slit gap, which is positioned immediately in front of the entrance slit (2). This reflects the majority of the heat from the entrance slit assembly to the outside of the instrument.

Fig. 9.5 Adjustable entrance slit assembly (Rondi)

The entrance slit assembly which is positioned at the focus of the telescope and the spectrograph main mirror is fully adjustable in rotation and allows up to 10 mm lateral movement (see Fig. 9.5).

From here the emerging beam reaches one side of the spectrograph mirror, where it is reflected as a parallel beam back towards the reflective grating. Note the extensive baffling. See Figs. 9.2 and 9.6.

The reflective grating, 63×72 mm, 1800 l/mm, was donated to Rondi by Leonard Wolff. The grating is firmly mounted in a rotational holder, the axis of rotation passing through the center of the front surface. Adjusting screws (Fig. 9.6, VR) allow for final adjustment and alignment of the assembly. The dispersed beam is then sent to the opposite side of the main mirror from the grating (see Fig. 9.2) where it is reflected back to the small mirror (3b) mounted in front of the camera attachment.

An external tangent arm (BR) and fine adjusting screw (M) fixed to the grating axis allows for the precise control of the grating rotation and wavelength selection. Also shown in Fig. 9.7 is the adjustable platform (F) which supports the digital imaging camera. The platform has fine micrometer control to allow precise focusing of spectral image onto the camera.

Rondi originally used a black and white modified webcam with a chip of 5.78×4.89 mm which was later replaced with a CMOS camera, PL-1M, this camera has a 1280×1024 array, pixel size 6.66×5.32 μm. The dispersion using this camera and the 1800 l/mm grating is 0.034 Å/pixel. This infers a resolution of approximately 0.13 Å, and R = 48000 (at Hα wavelength).

Fig. 9.6 Reflective grating arrangement (Rondi)

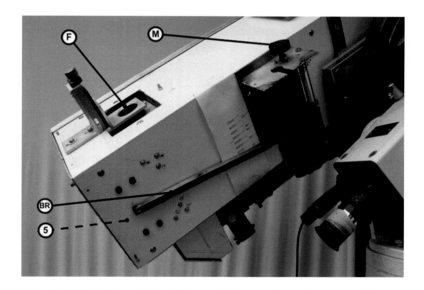

Fig. 9.7 Grating rotation arm and camera focus arrangement (Rondi)

9.1.1 Results

Figure 9.8 shows the same prominences recorded in Hα and He D3 wavelengths.

Rondi has also contributed examples of his excellent work on Ellerman bombs and magnetograms to Chap. 10.

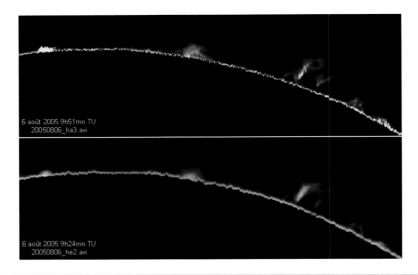

Fig. 9.8 Prominences recorded in Hα (*top*) and He D3 (*below*) (Rondi)

9.2 Rousselle

Philippe Rousselle has been an active SHG user for many years and was one of the first to make use of webcams.

His current digital SHG as shown in Fig. 9.9 makes use of a 90 mm 1296 mm focal length f14.4 achromatic objective and two 60 mm 900 focal length f15 doublet lenses in the spectrograph configured as a folded classical arrangement (Fig. 6.10). An 1800 l/mm grating gives a dispersion of 0.1 Å/pixel and a resolution of better than 0.5 Å.

The optics are mounted in a 230×230×1250 mm plywood box construction, with four internal sections—one for the telescope optics and two for the spectrograph optics.

The telescope presents a solar image 12 mm diameter to the entrance slit.

The adjustable entrance slit is constructed from two beveled blades screwed onto a pantograph aluminum support frame; this in turn sits on a large metal washer (Fig. 9.11) which allows rotation control.

Using a pair of achromatic lens with focal lengths close to 1 m, the chromatic focus position can vary up to 4 mm between the UV and Hα wavelengths. This problem

Fig. 9.9 General views of SHG (Rousselle)

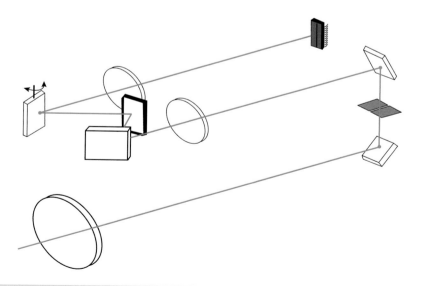

Fig. 9.10 Final revised optical layout (Rousselle)

applies to both the telescope and spectrograph optics, and requires a precision focusing mechanism on the entrance slit to satisfy both optical systems.

A compromise solution was used in which two first surface mirrors were mounted in a carriage assembly across the entrance slit and the whole assembly, mirrors and slit, mounted in a carrier which could be moved forward and back along the track for focusing as shown in Fig. 9.12.

Fig. 9.11 Grating rotation control and slit detail (Rousselle)

Fig. 9.12 Slit and folding mirror assembly on the axial track (Rousselle)

The telescope focus on the slit can then be adjusted for different wavelengths. This results in the spectrograph collimator working slightly away from the optimum position, but the loss of collimation has been found to be acceptable and a separate focuser mounted at the sensor compensates for this offset.

The grating used is unique, being larger than normal (62×72 mm), and having 1800 l/mm with two separate blazed areas. Two thirds of the surface is blazed to 2500 Å (UV) and remaining 1/3 at 6000 Å (close to Hα wavelength). See Fig. 9.13.

A fully adjustable grating support is mounted on a shaft (Fig. 9.13) and the grating assembly rotated using a tangent arm and a stepper motor drive. The drive mechanism is shown on the left hand side of Fig. 9.11.

The original design employed only one front surface mirror to fold the classical spectrograph arrangement. After initial testing, a third front surface mirror was introduced (see Figs. 9.10 and 9.14). The added mirror minimized the total angle and reduced the anamorphic factor between the collimator and the imaging lens.

A small flip mirror is mounted in the imaging path in front of the CCD to allow visual observation of the spectrum.

The imaging camera, visible at the right hand side of Fig. 9.15, is a Sony linear mono CCD, ILX751A which has an array of 2048×14 μm pixels. The software

Fig. 9.13 1800 l/mm grating in holder (Rousselle)

Fig. 9.14 Internal Layout (Rousselle)

Fig. 9.15 Slit assembly and Camera (Rousselle)

control allows exposures of between 10 and 136 ms to be used. At normal sidereal drift this gives over 1500 lines. The resolution obtained is 0.1 Å.

The SHG is mounted on a heavy DIY fork mounting with a smooth sector tape drive, which is very reminiscent of the Tuthill design (Sky and Telescope, July 1972, p. 47).

9.2.1   Results

The excellent results obtained in CaK and Hα are shown in Fig. 9.16.

Further results showing the Zeeman effect resolved with this SHG are shown in Chap. 10.

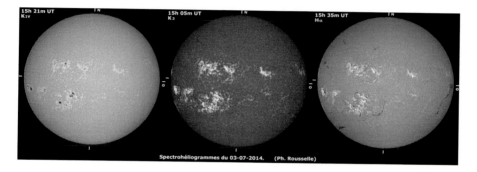

Fig. 9.16 SHG images: Ca II K1v (*left*), Ca II K3 (*center*), Hα (*right*) (Rousselle)

9.3   Poupeau

Jean-Jacques Poupeau designed his SHG along similar lines to Rondi's Ebert-Fastie design. His preference was to make use of reflective optics for the SHG to be free from the problems of chromatic aberrations and focus variations related to wavelength.

To achieve this design he used a Newtonian rather than a refractor telescope. He ground and polished the optics for the telescope himself. The primary mirror is 98 mm diameter, 1440 mm focal length (f14.7), mounted in a 120 mm diameter PVC tube which lies below the spectrograph parallel to the optical axis as shown in Fig. 9.17. To remove any diffraction effects associated with the normal Newtonian secondary supports, the secondary mirror is mounted on a BK7 window (seen in Fig. 9.18) which also encloses the telescope tube. The focal ratio was chosen to match the available mirror used in the spectrograph.

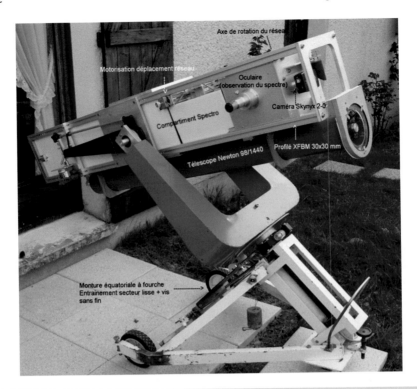

Fig. 9.17 General view of SHG, mounted on a home made fork equatorial mount (Poupeau)

The solar image is focused onto the slit assembly by axial movement of the primary mirror via a second front surface mirror (see Fig. 9.19).

The entrance slit is made from machined precision brass blades mounted on an aluminum cylinder (Fig. 9.20). This provides a 30 mm long gap and is usually set to a gap width of 15–17 μm. The small tangent arm and push screws provide for small angular corrections to the slit gap position.

The Ebert-Fastie spectrograph is designed around the primary mirror of an old Newtonian telescope, 210 mm diameter, focal length 1200 mm (f5.7), and housed in a rigid box frame constructed from 30 mm × 30 mm aluminum sections (XFBM). The walls of the box are removable and are made from lightweight sheets of 6 mm Depron foam—a closed cell polystyrene sheet—covered with black velvet. This makes assembly and modification very easy and allows the flexibility of adding accessories to the frame supports.

The spectrograph slit focus is achieved by axial movement of the main mirror (see Fig. 9.19).

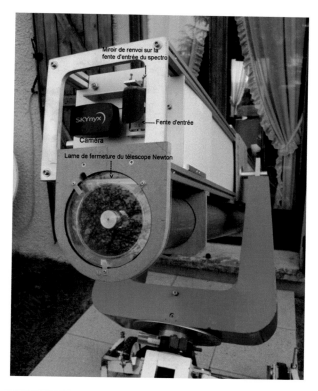

Fig. 9.18 Newtonian telescope showing secondary mirror supported from window (Poupeau)

The reflective grating employed is 58×58 mm, 1200 l/mm. The grating is located in a sealed box provided with a retractable window which provides a dust tight enclosure when the grating is not is use. Positioning the grating for the various imaging wavelengths is achieved using a tangent arm/lead screw arrangement driven by a stepper motor (see Fig. 9.21).

The recent addition of a reducer lens in front of the camera, an achromatic lens, 100 mm focal length giving ×0.5 magnification seen in Fig. 9.22, allows the full disk height spectrum to be recorded. It is located in the enlarged portion of the tube supporting the camera.

The original Skynyx 2 camera has been replaced by a Basler camera aca1600 g. The DIY fork equatorial mounting used was also recycled from a previous project.

Fig. 9.19 Newtonian focusing mechanism (Poupeau)

Fig. 9.20 Brass entrance slit jaws showing fine rotation mechanism (Poupeau)

Fig. 9.21 Top view of the SHG showing the grating tangent arm mechanism (Poupeau)

Fig. 9.22 Final SHG configuration (Poupeau)

9.3.1 Results

Poupeau processes his images using IRIS and applies the final enhancements with Photoshop CS3. Figures 9.23, 9.24 and 9.25 show some of the results obtainable with his SHG.

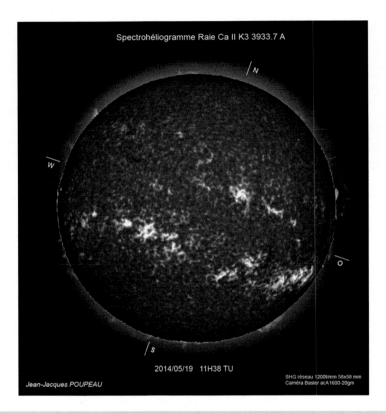

Fig. 9.23 SHG image taken in Ca II K3 (Poupeau)

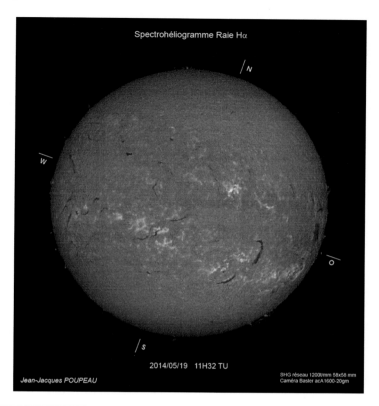

Fig. 9.24 SHG Hα image (Poupeau)

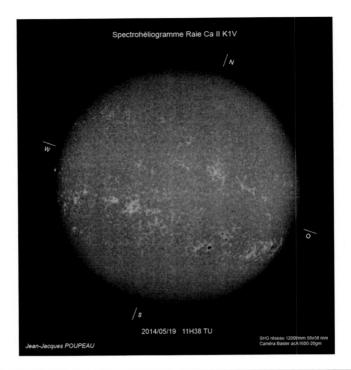

Fig. 9.25 SHG image recorded in Ca II K1v (Poupeau)

9.4   Defourneau

Daniel Defourneau was at the forefront of the digital SHG revolution. In 2001 he started to build his SHG.

His original design (Fig. 9.26) emphasized simplicity, using readily available materials. Defourneau made use of two small telescopes, a telephoto lens and a webcam mounted on board—salvaged from his closet shelves no less.

In the current arrangement an 80 mm achromatic objective, 800 mm focal length (f10) forms an image of the Sun on the entrance slit mounted in the focuser of a small 60 mm refractor, 350 mm focal length (f5.8). This acts as the collimator in a classical spectroscope configuration (see Fig. 9.27).

The slit jaws are cut, beveled and polished from pieces of brass sheet, and mounted at the end of a 32 mm diameter tube, slightly reduced in diameter to fit the standard 1.25″ focuser. The gap is set by eye to approximately 20 μm.

This arrangement (shown in Fig. 9.28) allows the tube and focuser to be used to precisely focus the solar image on the slit gap and rotation of the slit to align with the grating. The tube also acts as an aluminum radiator to dissipate heat.

Fig. 9.26 Original arrangement of the SHG. The *white* tube is the telescope (Defourneau)

Fig. 9.27 Current SHG general arrangement (Defourneau)

Fig. 9.28 Entrance slit (Defourneau)

Fig. 9.29 Gratings (Defourneau)

Over the years various gratings have been used with this basic set-up. Figure 9.29 shows the different grating options: 800, 1440, and 2400 l/mm. The 30 mm square 2400 l/mm was found to give the best results.

This grating is mounted on a rotational base centered on the optical axis in front of the collimator and fitted with a threaded rod which allows fine tuning of the grating rotation. The diffracted beam is directed to a 200 mm photographic lens which then images the spectrum into the webcam.

In June 2002 Defourneau wrote his SpecHelio software in Virtual Basic to process the spectral images from the webcam, discussed earlier in Sect. 8.5. As far as I can determine, this was the first time such a program was developed and used with the SHG. The initial images showed all the usual problems associated with the digital SHG.

Figure 9.30 shows one of the first digital SHG images ever obtained. This was recorded with a webcam at 10 frames per second (fps), with the solar disk drifting across the slit gap. The resulting BMP image was 2400 pixels wide, oval in shape and has the spectrum inclined, the wrong spectral lines, and transversalium due to defects in the slit. From this tentative beginning Defourneau refined his techniques to achieve spectacular results.

9.4.1   Results

The solar disk (Fig. 9.31) as recorded in Hα with the current instrument demonstrates the capability and performance of the SHG.

Fig. 9.30 Very early SHG image using a webcam and SpecHelio software (Defourneau)

Fig. 9.31 A Recent Hα image from the digital SHG (Defourneau)

Figure 9.32 shows the instrument and the schematic diagram of the SHG built by Peter Zetner. The instrument is constructed around a surplus Czerny–Turner monochromator housing. The original C-T optics were removed and only the grating holder assembly retained. Inlet and outlet ports were provided to support the collimator and imaging assemblies and a small front surface mirror directs the collimated beam to the grating in a folded classical design spectrograph.

Instead of the usual achromatic lenses Zetner makes use of the older M42 Pentax thread lenses, readily available on eBay. These lenses are often well corrected for aberrations, notably the Pentax Takumars, and Olympus Zuiko. To keep the overall size of the instrument as small as possible, an "Astro Rubinar" f10 lens (100 mm aperture), of catadioptric Maksutov design is used as the telescope. This is fitted to a UV-IR blocking filter (Schneider True-cut, 100 mm × 100 mm) as an ERF to give protection to the built-in cemented achromats located near the focal plane of the lens. The telescope produces a solar image of approximately 9.3 mm in diameter which is focused onto a 25 μm × 18 mm long slit (a recovered 18 mm high MonoSpec Bayonet slit for Jarrell Ash spectrometers). This is held in place by a modified version of the original monochromators slit holder. The grating holder is rotated by a simple axle and knob.

A collimator telelens with focal lengths ranging from 300 to 750 mm has been used in various versions of the instrument. Used with the Takumar 300 mm focal length lens, the collimator matches the f ratio of the Rubinar and the 30 mm square grating (2400 l/mm) used.

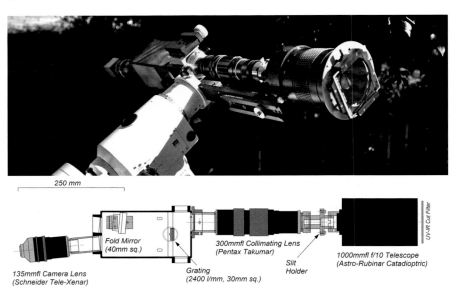

Fig. 9.32 General mechanical and assembly layout mounted on an NEQ6 mount (Zetner)

The final diameter of the solar image at the camera is determined by the imaging lens, a Schneider 135 mm telelens. This lens gives a reduction magnification of 135/300, ×0.45. The solar image size at the camera is therefore 4.2 mm and the effective slit gap becomes (25×0.45), 11.3 μm. This equals a 2.4 pixel width, both of which are a good match for the DMK31AU03 sensor size (5.80 mm×4.92 mm, 4.65 μm) being used for imaging.

This set-up provides a spectral resolving power of approximately R = 20000 with the 25 μm entrance slit and a 2400 l/mm grating.

The SHG is usually mounted on a Skywatcher NEQ6 Equatorial mount.

Zetner produces his images with the standard drift scan, capturing at 15 fps. The resulting 1920 frame AVI file is processed with VDub and ImageJ before final enhancement using Photoshop.

9.5.1   Results

Figure 9.33 shows a comparison of disks imaged at a resolution of 0.25 Å in Hα and He D3 (5876 Å) on 15 July 2014. Entirely different aspects of the solar atmosphere are visible. For example, active regions in Hα show plage features that correlate directly with similar regions of "black plage" in He D3 wavelength. The filaments in Hα correlate with similar features in He D3, in both cases visible as dark absorption features. Further examples of Zetner's work are shown in Chap. 10.

Fig. 9.33 SHG results. Hα (*left*), He D3 (*right*) (Zetner)

9.6 Smith

All of the previous examples of the SHG have been built using reflection gratings. Douglas Smith has gone against the trend and constructed his SHG using a transmission grating arrangement. This is shown in Fig. 9.34. Typically, reflection gratings are more efficient than transmission gratings. As the grating density (l/mm) increases, the transmission gratings lose efficiency particularly at longer wavelengths (see Fig. 9.35). With the amount of light available to solar observers, this is not a major problem.

Small transmission grating spectroscopes are readily available to the amateur astronomer from manufacturers like Paton Hawksley, Elliot Instruments and Shelyak. These spectroscopes are compact yet work extremely well. The reason these small spectroscopes are so compact is that they use a grism (see Sect. 4.6). Typically a 600 l/mm transmission grating is attached to a 38° angle prism. In front of the grism is a 23/50 μm slit followed by a collimating doublet.

Smith thought that it should be possible to build a SHG using a similar approach. The transmission spectroscopes are designed to view the maximum amount of the visible spectrum; with the prism part of the grism having an angle such that that green light (5000 Å) is undeviated (i.e. the middle of the spectrum image sits on the optical axis). For a SHG, we want to be able to zoom into a fairly narrow region of the spectrum and ideally this target wavelength should be on axis. To achieve this, Smith devised a variable grism. Basically, instead of putting a transmission grating on the angled surface of a prism, a flat transmission grating is sandwiched between two prisms (see Fig. 9.36).

Fig. 9.34 Transmission grating SHG (Smith)

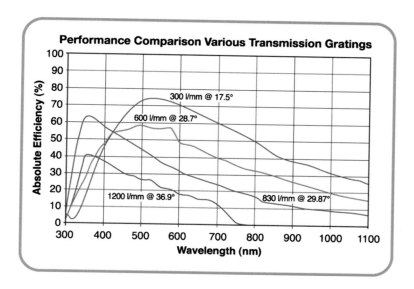

Fig. 9.35 Efficiency curves for transmission gratings (Thorlabs)

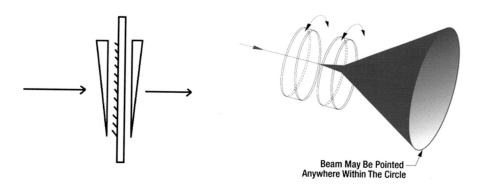

Fig. 9.36 Double prism grating combination (Smith/Thorlabs)

He calls this configuration an "Amici grating", an analogy to the well-known double Amici prism configuration. Using the SimSpec spreadsheet (See Appendix B), it was calculated that putting 10° prisms in front and back of a 600 l/mm transmission grating would leave 5800 Å green light undeviated. To bring a target wavelength towards the optical axis, instead of using a single prism in front and back of the grating, a pair of prisms that could be rotated on their axis was used. This is called in laser optics a "beam steering" configuration (Fig. 9.36). Collimated light sent through a pair of rotatable wedge prisms can be made to trace out a solid cone, and by using a fixed 10° prism and a rotating 4° prism an effectively variable prism between 6 and 14° is created.

Using SimSpec again, it was calculated that, using a 600 l/mm grating in conjunction with a 6° prism provides undeviated light in the violet in the first diffraction order. By increasing the effective prism angle, the center wavelength can be increased up through the infrared and all the way to violet in the second diffraction order at 14°. When tested by Smith this approach worked very well. The wedge prisms are rotated manually inside Thorlab's one inch lens tubes. But ideally levers should be added, rather like in an atmosphere dispersion compensator (see, for example, the Pierro-Astro ADC).

To increase the spectral resolution as much as possible, the transmission grating was upgraded to a 1200 l/mm, 25 mm×25 mm in size (Thorlabs part number GT25-12). The substrate is 3 mm thick Schott B270 glass. The grating is blazed with a 36.9° groove angle. According to Thorlabs, the grating has a transmission of 37 % at 4000 Å and 12 % at 7000 Å. While this relatively low efficiency would not be suitable for night-time astronomy, for solar work it is not a problem.

Smith's preference was to use very high quality fixed slits made using laser cutting of stainless steel giving the narrowest slit possible and also good uniformity as possible across its length, to minimize transversalium. A 5 μm, 3 mm long fixed vertical slit from Thorlabs (part number S5R) was selected.

To fully illuminate the 25 mm grating, a 100 mm f4 telephoto M42 camera lens was used as a collimator. The lens' full open aperture of 25 mm matches the size of the grating. The slit must be placed at the lens focus. For M42 Pentax lenses, by far the most common lenses and the easiest ones to use, the flange focal distance is 45.46 mm. The focusing mechanism on the lens is used to exactly put the slit at the correct focal point. Once collimated correctly, this lens should not be adjusted again.

Another M42 lens, this time a Pentax 150 mm f4, was used as the imaging lens to magnify and focus the light from the grism onto a camera. The 37.5 mm aperture, slightly larger than the grating, is required to capture all the diffracted rays.

The longer focal length imaging lens gives a system magnification of ×1.5 and an apparent slit gap of 7.5 μm at the CCD. To connect the camera lens, a camera bellows made by BPM in the UK was added. This allows some extra rotational degree of freedom for placing the desired spectral line near the center of the camera sensor (see Fig. 9.37).

An additional Pentax 200 mm f4 lens is employed as the telescope part of the SHG to focus an image of the Sun onto the slit. The resulting 50 mm aperture is

Fig. 9.37 Layout of the transmission grating spectrograph (Smith)

adequate and the 200 mm focal length gives a solar image of almost 2 mm on the slit, allowing full disk imaging in a single scan.

The camera used is a ZWO ASI 120MM camera, a modestly priced mono CMOS camera with 3.75 μm pixel and a USB2.0 interface, with good sensitivity across the visible spectrum (Smith comments that it actually seems to be much better in the violet than some CCD cameras). The sensor size is 1280×960 pixels and at full resolution the camera is specified at 35 fps. By using the image capture program FireCapture, a smaller region of interest (ROI) can be selected which increases the fps considerably. 300 fps seems to be the stable upper limit which the USB2.0 version will work. The new USB3.0 version will presumably go even faster! The reason high speed is important is that it allows fast scanning of the SHG across the Sun so that the image is acquired in as short a time as possible.

As Fig. 9.34 shows, the SHG is quite small and light. The whole apparatus can be held in one hand and is about the same size and weight as a PST. This allows it to be used on a fairly small telescope mount, in this case a Vixen GP2. The mount is aligned roughly north and sidereal tracking is turned on. The manual scanning speed of the mount goes up to a maximum of 32× sidereal rate. At 1× sidereal rate (i.e. if the tracking were turned off), the Sun would scan across the SHG slit in 120 s. Using 32× sidereal rate scanning, the SHG slit only takes 3.75 s to move across the Sun. Doing four complete scans in 25 s works out quite well. This generates a reasonable amount of data all at once but the file size does not become too large, remaining less than 1Gb.

The CaK image strips shown in Fig. 9.38 were extracted from an AVI obtained using a ROI of 816×128 pixels. 7685 frames were captured in four scans over 25 s (average fps of 307). The maximum height of the solar image is approximately

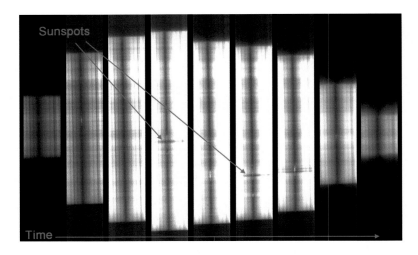

Fig. 9.38 Sample images from AVI scan (Smith)

720 pixels. In a typical full disk 3.75 s scan about 1150 frames were captured. This number becomes the number of pixels wide the Sun appears when processed, the aspect ratio being 1150/720 = ×1.6. The images were processed using VirtualDub and SpecLineMerge.

9.6.1 Results

For comparison, images obtained using two different spectral bandwidths are shown, 1 pixel wide and 10 pixels wide. See Fig. 9.39.

As the current instrument dispersion is approximately 0.2 Å per pixel, this corresponds to a resolution of about 0.8 and 2 Å,—similar to the range of the commercial Hα and CaK solar filters. After processing, the images compare fairly well with a PST CaK image taken the previous day.

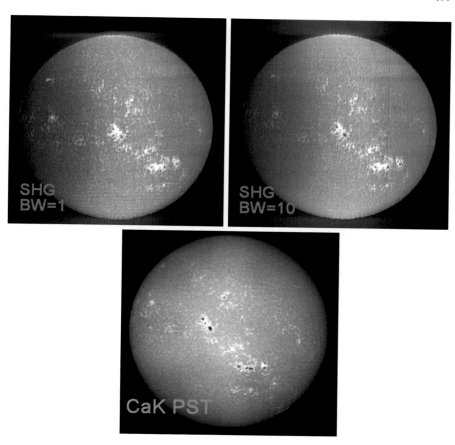

Fig. 9.39 Comparison SHG CaK images and PST CaK image (Smith)

9.7 Mete

Italian Fulvio Mete has been active in spectroscopy for last 17 years and has designed and built many spectroscopes. He recently developed his HIRSS as a SHG and later upgraded to his HIRSS2 and more recently to his VHIRSS (Very High Resolution Solar Spectroscope) design.

The VHIRSS is a compact and lightweight Littrow style SHG. It uses a 62 mm, 480 mm focal length (f7.7) refractor telescope, adjustable slit (Surplus Shed) and an 80 mm, 600 mm focal length (f7.5) ED refractor as the Littrow spectrograph. A 50 × 50 mm 2400 l/mm holographic grating is mounted in front of the objective. The overall length is 1200 mm and weight 8 kg. The SHG is then fully portable and easily installable on a medium size mount (CG11, NEQ6 etc.).

Fig. 9.40 The VHIRSS SHG mounted on a G11 EQ mount (Mete)

Figure 9.40 shows the current configuration of the VHIRSS. The solar image produced by the 62 mm aperture telescope is 4.8 mm diameter and fits neatly within the Surplus Shed slit gap height. In this latest incarnation of the VHIRSS the telescope has been mounted on a dovetail arrangement with a fine screw focusing adjustment (see Fig. 9.41).

Mete has mounted the slit assembly slightly off axis and the pick off mirror at a similar distance on the opposite side to further reduce the Littrow total angle and the astigmatism.

The grating is held in a DIY plastic cell which is locked, using two nylon screws, to a 5 mm threaded shaft positioned centrally across the dew cap of the ED80 telescope used as the Littrow collimator (see Fig. 9.42).

Although this arrangement is not ideal—the axis is not aligned through the grating surface, in this application it works very well. The grating adjustment is achieved using an external tangent arm and screw, as shown in Fig. 9.43.

The long focal length of the Littrow collimator gives a high dispersion, 0.029 Å/pixel at Hα and 0.02 Å/pixel at CaK when using the DMK41 camera.

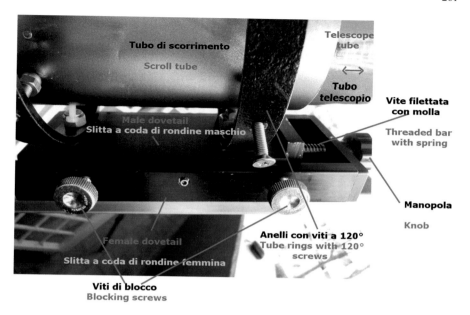

Fig. 9.41 Focuser adjustment on the VHIRSS telescope (Mete)

Fig. 9.42 PVC Grating holder and nylon axle locking screws (Mete)

Fig. 9.43 Grating holder and adjustment arm on the ED80 dew shield (Mete)

9.7.1 Results

Figures 9.44 and 9.45 show the capabilities of the HIRSS2 and VHISS SHG configurations. Mete believes he was the first to make use of the "X Binning" synthetic flats (see Sect. 5.2) and his results demonstrate the ease of its use.

The simplicity and outstanding performance of the VHIRSS design has proven very popular with other amateurs.

The first amateur digital SHGs in China were built from the VHIRSS concept by Wah-Heung Yeun and Man-Fai Law (Fig. 9.46a, b).

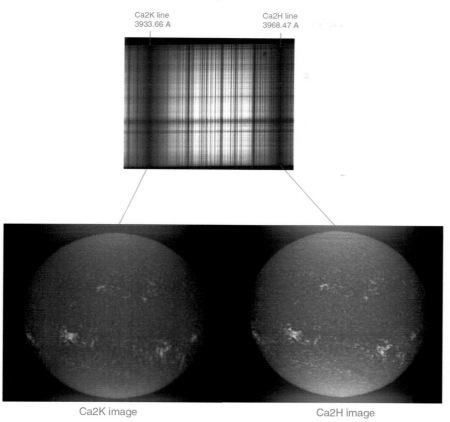

Fig. 9.44 Sample images from the original HIRSS2 (Mete)

AR 1476 →

Sun Ha May 12 2012 h 8.10 UT
VHIRSS Spectrohelioscope mode
Fulvio Mete
Rome, Italy

Fig. 9.45 VHIRSS Hα image (Mete)

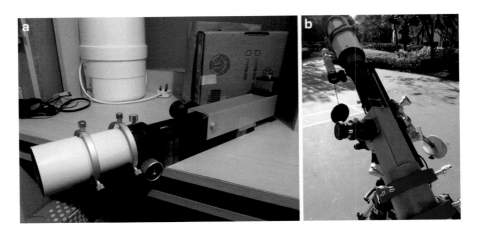

Fig. 9.46 (**a**) VHIRSS by Wah-Heung Yeun (**b**) VHIRSS by Man-Fai Law (Mete)

9.8   Chagniard

The SHG constructed by Vincent Chagniard working in France was inspired by Daniel Defourneau's work. The telescope utilizes the optics of an old 60 mm, 900 mm focal length (f15) refractor (Fig. 9.47) mounted in a PVC tube. The tube slides in the mounting rings to allow focusing of the solar image onto the adjustable slit (Surplus Shed) and produces a solar disk approximately 8.33 mm diameter.

Chagniard has constructed a classical spectrograph using standard photographic lenses. The adjustable slit is mounted into a plastic camera body cover and sits on a 300 mm photographic collimating lens (see Fig. 9.48).

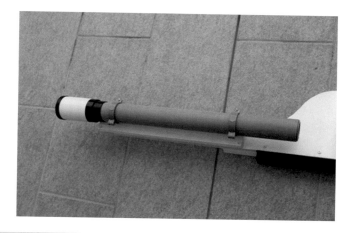

Fig. 9.47 Telescope section of the digital SHG (Chagniard)

Fig. 9.48 Surplus Shed Slit mounted in a lens body cap (Chagniard)

Figure 9.49 shows the arrangement of the two telelenses and the grating within the spectrograph housing. At the upper right is the 300 mm focal length collimator fitted with the entrance slit. The grating, held in the Eksma mount sits quite far from the collimator and the imaging lens, a70-210 mm focal length zoom lens (used at 135 mm focal length) seen below the collimator. The total angle between the collimator and imaging lens was selected at 15° to minimize the anamorphic factor (Sect. 6.6) and as a compromise to the total width of the instrument.

The reflective grating used is a blazed Optometrics 30×30 mm, 2400 l/mm, which matches well with the telescope f ratio and collimator focal length. This grating is mounted securely on an Eksma Optics square optics holder (Fig. 9.50).

The M6 mounting screw on this holder is unfortunately not aligned with the front surface of the grating, but for the limited fixed positions used (for Hα and CaK) this doesn't pose a great issue. The grating angle is adjusted by hand to center the required wavelength in the camera. The use of standard camera lenses although adding some weight, simplifies the focusing of the spectrograph.

Figure 9.51 shows the early results obtained with the SHG. The completed instrument is shown in Figs. 9.52 and 9.53.

The SHG is currently mounted on a Vixen GP Equatorial mount. Chagniard mentions that a larger mount would be preferred as wind can be an issue causing vibration (see Fig. 9.53).

Fig. 9.49 Spectrograph layout. The grating at the left and the collimating lens with entrance slit at the *top right*. The imaging lens is below with a total angle between them of approximately 15° (Chagniard)

Fig. 9.50 Eksma Grating holders (Chagniard)

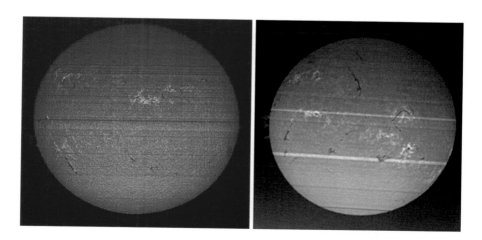

Fig. 9.51 Early Hα results showing the typical transversalium features (Chagniard)

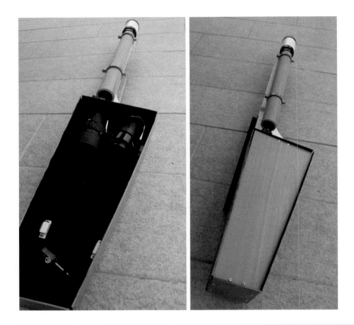

Fig. 9.52 General arrangement of the SHG (Chagniard)

Fig. 9.53 SHG seen here seen open, mounted on a Vixen EQ mount (Chagniard)

In this configuration the SHG gives a theoretical dispersion of 17 Å/mm at Hα and 25 Å/mm in CaK. The DMK31 camera used (4.65 μm pixel) gives 0.08 Å/pixel and 0.11 Å/pixel for Hα and CaK respectively. This allows recording of Doppler shifts of 3.6 km/s/pixel at Hα and 8.6 km/s/pixel at CaK.

Chagniard uses FireCapture for imaging acquisition and SpecLineMerge/Gimp for final processing.

9.8.1   Results

The following images, Figs. 9.54, 9.55 and 9.56 show the typical features recorded during the scan in Hα.

Some of the current full disk images being obtained at various wavelengths are shown in Figs. 9.57 and 9.58.

Fig. 9.54 Hα line showing the passage of a sunspot (Chagniard)

Fig. 9.55 Hα line showing emission feature (Doppler shift) (Chagniard)

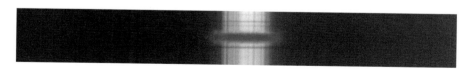

Fig. 9.56 Hα prominence at the extreme solar limb (Chagniard)

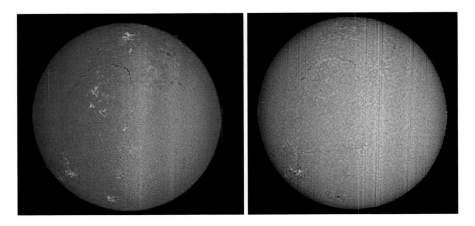

Fig. 9.57 SHG images in Hα (*left*) and Hβ (*right*) (Chagniard)

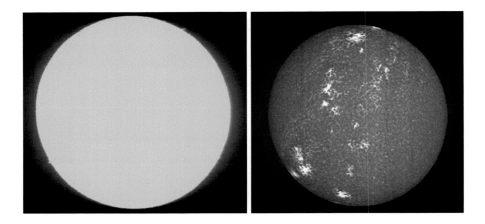

Fig. 9.58 SHG images in He D3 (*left*), CaK (*right*) (Chagniard)

9.9   Bartolick

For the past 48 years, Joe Bartolick has been constructing SHG/SHS' from his home in California. The latest instrument rebuilt in 2009 and upgraded in 2013 is detailed here. Over the years the usage has ranged from visual (in a SHS configuration) through to imaging on photographic film and digital detectors such as linear diode arrays and CCDs. The current instrument, originally configured as a SHS, now makes use of a modified business card line scanner as the detector.

Fig. 9.59 View of the SHG assembly showing the cœlostat (*left*) and the main SHG housing (*right*) (Bartolick)

The instrument itself still retains many of the features incorporated in its original visual configuration, so is of interest to both SHS and today's digital SHG builders. The early heritage is immediately obvious from both the physical size of the instrument and the fact that it is fed sunlight via a cœlostat as seen in Fig. 9.59.

A two mirror cœlostat with 100 mm front surface mirrors shown in Fig. 9.60 provides the solar beam to the telescope section. The cœlostat is driven by an AC synchronous motor through a worm reduction. The primary mirror mount is fixed at the observer's latitude, and slides north-south in tracks to allow for solar declination changes. There are three of these tracks to allow for east-west movement, both to avoid the shadow of the secondary mirror and to allow better mirror fill at large hour angles late in the day. The secondary mirror is mounted in a fork with motors driving it in both axis, these are connected to DC amplifiers that get their signals from a quadrilateral silicon detector. The detector is mounted on a mechanical stage that would ordinarily be used on a microscope. This allows for offset guiding. An image of the Sun is formed on the detector by sampling a 10 mm square area in the center of the objective lens.

The guide system corrects for poor polar alignment or other drifts in image position, effectively maintaining the solar image position on the entrance slit. The guider is disabled during scanning.

The telescope optics, an 81 mm 1215 mm focal length (f15) achromatic objective, is folded using a prism (Fig. 9.61, Turning Prism) and the solar image enlarged by a pair of lenses.

Fig. 9.60 Cœlostat arrangement—SHG housing in background (Bartolick)

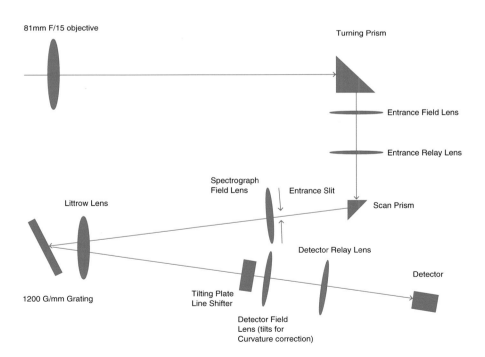

Fig. 9.61 Optical layout of the SHG. Note the use of field lenses (Bartolick)

The entrance lens system shown in Fig. 9.61 is a combination of the Entrance field lens (500 mm focal length) and Entrance relay lens (a 75 mm focal length camera lens). This basically forms a 90 mm eyepiece, providing the amplification to give an effective focal ratio of f28 and projecting an enlarged solar image (22 mm diameter) onto the entrance slit through the scan prism. The exit pupil is positioned to lie on the scan prism. This ensures uniform illumination of the grating regardless of scan angle.

In the original SHS configuration this scan prism played an important role. It effectively rotates at a constant rate through a large enough rotational angle to move the solar disk across the slit gap. The drive mechanism used for the scanning will be discussed later.

The entrance slit assembly has jaws made from tool steel, 30 mm long and mounted in an aluminum block. The entrance slit gap is fixed at 25 μm.

A field lens is positioned 11 mm past the slit to present an exit pupil onto the Littrow lens. This provides uniform illumination and reduces vignetting effects in the spectrograph. The Littrow lens is a 60 mm diameter, 1270 mm focal length (f21) achromat mounted on a micrometer controlled slide for precise focusing. The reflective grating used is the Thorlabs 50 mm square, 1200 l/mm, blazed at 5000 Å, and sits on a sine bar driven mount calibrated to give direct wavelength readout (see Fig. 9.62).

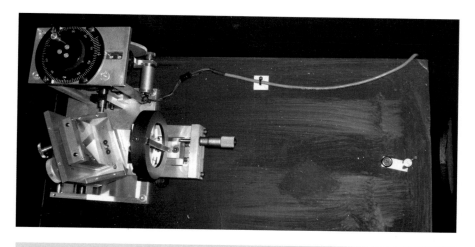

Fig. 9.62 Grating holder and Littrow lens—showing wavelength readout and stops in optical path (Bartolick)

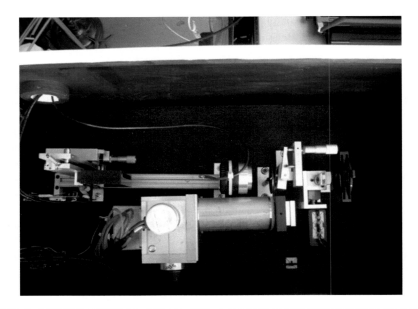

Fig. 9.63 Scanner and detector area. The scanner and entrance slit are at the bottom; at the top are the parts of the detector assembly. From the right are the ND filter, line shifter, tilting field lens, mask, relay lens and detector (Bartolick)

The diffracted light is then reimaged by the Littrow lens. In the imaging system a filter holder (for ND filters) and an aperture mask with a 2 mm slot restrict the final spectral beam. This is followed by an 8.5 mm thick glass block—the tilting line shifter. This is rotated by a galvanometer which allows fine control of the wavelength being observed (±1.5 Å) (useful for Doppler measurements and checking on the line center position).

Following the line shifter there is a further relay lens system. See Fig. 9.63.

The relay lens images the Hα line onto the detector with a magnification of ×4.45. The first optical element of the relay system acts as another field lens which has two functions. First, it images the Littrow lens onto the camera lens and ensures uniform field illumination. The second function is very innovative! The field lens can be rotated about a vertical axis via a micrometer and sine arm arrangement. The purpose of this is to eliminate the curvature of the spectral lines at the spectrograph's final focus, i.e. the smile effect. Tilting the field lens has the same impact as tilting the parallel glass plate of the line shifter, except that the field lens is plano-convex and thus thicker at the center. This causes the shift at the center to be greater than the shift at the edges, resulting in a straightening of the lines. As the field lens is located almost at the focal plane the astigmatism introduced is too small to be detected in the final image. The tilted field lens can be set to correct for any wavelength.

The final element in the optical path is a Nikkon 50 mm El-Nikkor enlarging lens. The combination of these lenses, gives a magnification of ×4.45 and the effective final focal length of the spectrograph imaging system therefore becomes 5640 mm, which results in a 53 mm high spectral image at the sensor.

All of the features described above were built into the original SHS to provide a suitable final image size for visual observing, but they still prove useful in the digital application.

Returning to the scanning prism drive system; the design is highly influenced by the linear CCD used by Bartolick for imaging. This is a 1275 pixel linear CCD with 42 μm pixels. The chip height is 53.55 mm. The detector assembly, complete with electronics, was removed from a Cardiris 4 monochrome business card scanner originally designed to scan a business card in 6 s. An electronic interface was constructed, replacing the original paper detect switch with a transistor closure. This provides a signal to the scanner electronics to effectively replicate the card scanning process.

When the scanner is commanded to start, the 5 V pulses that originally drove the LED illumination bar now are used to increment a 12 bit counter, which drives a 12 bit analog to digital converter. The output from this in turn drives the galvanometer that indexes the position the scanning table which results in the rotation of the prism, giving a full disk scan across the entrance slit in 6 s.

After a set number of lines are scanned (2125 in the 6 s scan) the paper detect closure is opened and the counter reset. The scanner software then thinks a card has passed through, and sends the image to the computer via USB. The scanner is controlled internally by a factory programmed IC which unfortunately cannot be reprogrammed. The end result is an 8 bit image, 1275 by 2125 pixels.

One of the interesting features incorporated into the SHG is the use of blocking strips and stops in the Littrow spectrograph. Imaging the bright solar disk results in unwanted reflections from the lens surfaces and can lead to a significant reduction in image contrast.

A blocking strip is positioned immediately in front of the Littrow lens (A) to suppress the bright point reflection from the first surface and a small plastic rectangle on a wire support (B) suppresses and traps reflections from the rear surfaces of the Littrow lens. These are shown in Fig. 9.62 and explained in the diagram, Fig. 9.64.

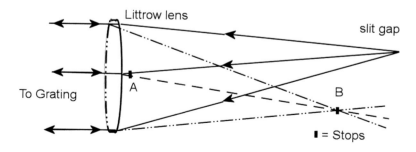

Fig. 9.64 Suggested stop positions to suppress Littrow lens reflections (after Bartolick)

Bartolick has recently built a new detector, using a Kodak KLI-2113, 2098 by 3 tri-linear CCD and is currently working on the software to integrate it into the SHG. The image scale achieved is 1.7 arc sec per pixel and the measured spectral resolution being obtained is 0.34 Å, at Hα.

9.9.1   Results

Figure 9.65 shows the quality of the results being obtained. The small white box shown (top right hand corner) is 10 μm square.

Figure 9.66 shows a Hα image obtained with the digital SHG compared to an image from the BBSO patrol telescope which used a 100 mm objective diameter, a 1K by 1K CCD and a Zeiss 0.2 Å Halle filter. An additional example of the high definition results being achieved by Bartolick is also shown in Chap. 10 to follow.

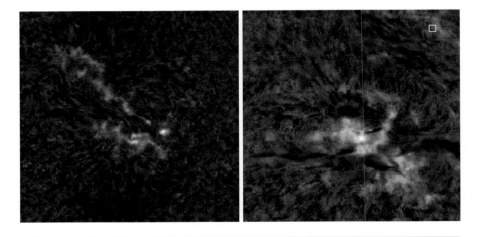

Fig. 9.65 Hα Images demonstrating the high resolution achieved (Bartolick)

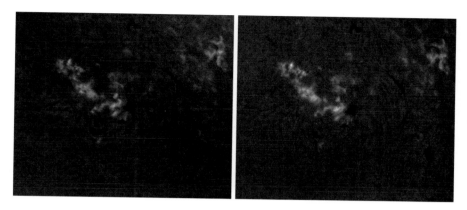

Fig. 9.66 Hα image (*left*) compared with BBSO (*right*) (Bartolick)

Webpages

http://www.astrosurf.com/rondi/obs/shg/index.htm
http://www.alldatasheet.com/datasheet-pdf/pdf/89520/SONY/ILX751B.html
http://www.astrosurf.com/spectrohelio/index-en.php
http://uk.rs-online.com/web/p/tubing-struts/2899054/
http://eksmaoptics.com/opto-mechanical-components/optical-mounts-830/rectangular-
    optics-holder-830-0100-830-0101-830-0110-830-0111/
http://www.apm-telescopes.de/en/optical-accessories/flattener-reducer-correctors/apm-pierro-
    astro-adc-atmospheric-corrector-t2.html
http://www.lightfrominfinity.org/VHIRSS/VHIRSS.htm
http://www.lightfrominfinity.org/Hirss2%20spettroelioscopio/hirss2_spettroelioscopio.htm
http://adsabs.harvard.edu/abs/2013AsUAI...1...22 M
http://www.baslerweb.com/en/products/area-scan-cameras/ace/aca1600-20gm
http://www.macrobellows.com/bpm_accessories.php

Chapter 10

SHG Images and Science Achievements

The current SHG users have already demonstrated some of the excellent results being obtained from their instruments, which were covered in the last chapter. In this chapter we show some of the exciting images and science which can only be achieved with the SHG at different wavelengths and bandwidths. One of the strengths of the SHG is that we can accurately position the acquisition bandwidth at a nominated wavelength and known distances (in Å) from the absorption line core CWL, something difficult to assess with normal commercial filters.

This flexibility of applying varying bandwidths in the SHG mosaics allows us to image the chromosphere at different heights and then compare the images from different parts of the chromosphere. We can also determine Doppler shifts, Magnetic fields in and around sunspot groups and more.

10.1 On Band, Red and Blue Hα Wings

Figure 10.1 shows the dramatic changes seen as the central wavelength is moved across the Hα line. On band the filament and structure is quite pronounced as is the plage area (left hand image, center right). As we start to move into the red wing (+0.45 Å) the Doppler shift reduces the visible filament structure and the plage is lost. Into the red wing (+0.78 Å) we are seeing the lower regions of the chromosphere close to the photospheric boundary with the flocculi visible in the background.

© Springer International Publishing Switzerland 2016
K.M. Harrison, *Imaging Sunlight Using a Digital Spectroheliograph*,
The Patrick Moore Practical Astronomy Series, DOI 10.1007/978-3-319-24874-5_10

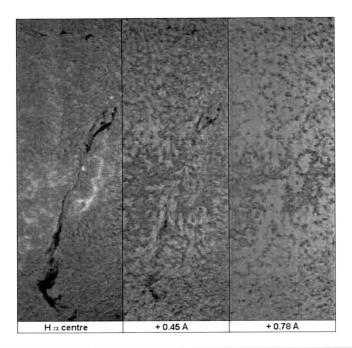

| Hα centre | + 0.45 A | + 0.78 A |

10.2   Ellerman Bombs (Moustaches)

Rondi captured the Ellerman bombs shown in Fig. 10.2. The two bright spots indicated are the transient Ellerman bombs. Note the bright "moustache" emission lines running across the associated spectral images from approximately ±1 to ±4 Å from the Hα centerline. It was the absence of emission structure in the Hα core region which caused Severny to call these features moustaches.

Fig. 10.2 Emission moustaches from Ellerman bombs (Rondi)

10.3   CaH, H1 and H3 Regions

As illustrated in Fig. 10.3, smaller differences in wavelength can also make large differences in the appearance of the solar disk. The figure shows comparison images taken (3 July 2014) only 0.5 Å apart in the broad Ca H line near 3968 Å at a resolution of less than 0.3 Å.

At the center of the CaH line (Ca H3), the image shows the upper regions of the chromosphere and the chromospheric network exclusively, whereas only 0.5 Å away (Ca H1) lower, photospheric features (sunspots) become more visible.

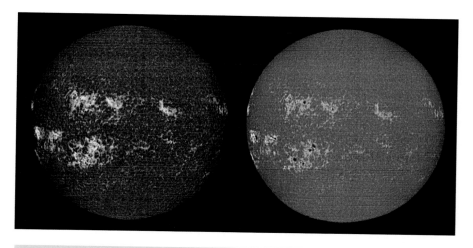

Fig. 10.3 CaH images showing the difference 0.5 Å off band (Zetner)

10.4 Doppler Shift Images

The digital SHG has the ability to produce calibrated images which can then provide the data for detailed study of velocities of solar features, magnetic field effects and more. By employing the Doppler shift calculations to the data the results allow the amateur to construct some extremely interesting illustrations of the dynamic material movements as they occur. For example, Fig. 10.4 shows a "dopplergram" of a prominence on the eastern solar limb (21 June 2014), captured in CaH light. The three color images show a velocity map of the prominence, highlighting regions of prominence plasma moving toward us (blue coloring) or away from us (red coloring) at the indicated velocities with respect to the CaH line center. Such maps are relatively easy to construct by producing spectroheliograms at known wavelengths offset to the blue and red sides of a spectral line center increasing by fixed increments and then subtracting the corresponding images in pairs.

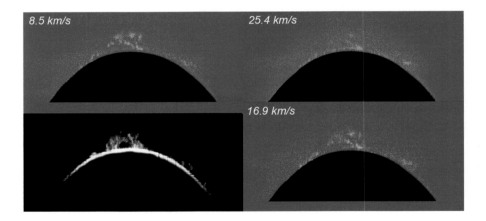

Fig. 10.4 Dopplergram of a prominence (Zetner)

10.5   Filaments/Prominences

The wealth of detail which can be recorded in Hα wavelengths is well illustrated in Fig. 10.5. The background mottling shows the evidence of magnetic streaming; the active area near center clearly shows the darker overlying filaments and the bright plage areas surrounding the sunspot. Towards the solar limb on the upper left the filament is seen to change into a significant prominence or filaprom which appears to sit in front of a larger quiescent hedgerow type prom. Also note the spicule "sierra" along the limb edge. The quality of this image emphatically demonstrates the imaging capabilities of the digital SHG.

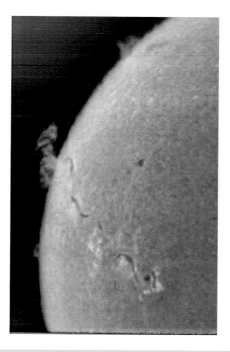

Fig. 10.5 Filaments, filaproms, plage and spicules (Bartolick)

10.6   Zeeman Magnetic Images

The splitting of the FeI line at 6302.5 Å due to the Zeeman effect from the underlying sunspot AR 1092 is visible in Fig. 10.6, where the penumbra of the sunspot is seen as a horizontal grey band on either side of the darker umbra area band.

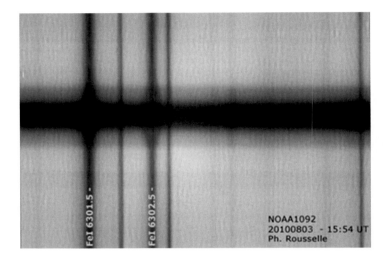

Fig. 10.6 Zeeman effect at FeI (6302.2 Å) (Rousselle)

10.7   Magnetography and Magnetograms

Figure 10.7 shows a magnetogram of the active area AR0905 produced using the FeI (6302.2 Å) line and using the techniques outlined in Sect. 5.4.3. The comparison image from the MDI on the SOHO satellite confirms the quality of the results being obtained.

The images seen here only represent a small sample of the scientific results capable of being obtained by the SHG. There are tremendous opportunities for ongoing scientific work for any dedicated SHG user.

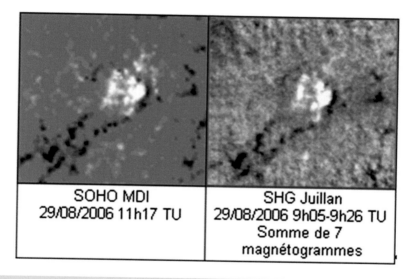

Fig. 10.7 FeI (6302.2 Å) magnetogram AR0905 (Rondi)

10.8   Kasi Solar Imaging Spectrograph (KSIS)

The KSIS system developed by SOS in Korea and prototyped as the FISS system on the Big Bear solar telescope I think gives some indication of the future capabilities of the digital SHG.

Figure 10.8 shows comparison Hα and CaII images of the active area AR11305. Each panel has a field of view of 40 arc seconds (horizontal) and 60 arc seconds (vertical). The rows (a–e) at top and (f–j) below show scans at −1 Å, −0.6 A, CWL, +0.4 Å and +0.9 Å from the Hα and CaII (8542 Å) wavelengths.

The KSIS system shows the dramatic images (and scientific work) which can be achieved with ground based instrumentation.

Wavelength scan images of Hα

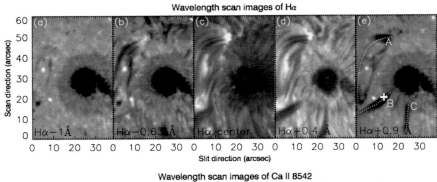

Wavelength scan images of Ca II 8542

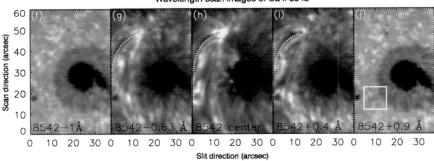

Fig. 10.8 AR 11305 in Hα (*upper row*) and Ca II 8542 Å (*lower row*) at different wavelengths. Features *A–C* marked in panel (**e**) denote (*A*): a closed loop that showed relatively horizontal flows, (*B, C*): apparently open loops that showed only downflow motions in Hα. *Square*: the same loop as *B*, but visible in Ca II 8542 Å as a faint and *dark* structure in the *middle* of the *square* (SOS)

Further Reading

Chae, J. (Ed.): Initial results from the fast imaging solar spectrograph (FISS). Springer (2015)

Webpages

http://www.astrosurf.com/rondi/obs/shg/magnetogram.htm
https://en.wikipedia.org/wiki/Waveplate#Quarter-wave_plate
http://www.thorlabs.hk/newgrouppage9.cfm?objectgroup_ID=7234
http://www.comaroptics.com/products/filters/retarders/mica-retarders

Chapter 11

The Spectrohelioscope (SHS)

The successful development of the digital SHG has been built on a long history of spectroscopy, spectroheliographs (SHG) and visual spectrohelioscopes (SHS) as discussed in Chap. 4. In this chapter we will briefly review the current status of the visual SHS as implemented by amateurs. It's fair to say that the complexity of the SHS, and the large focal lengths required to obtain acceptable visual results has worked against the ongoing development of the SHS. Fred Veio's construction notes for the SHS over the past 30 years have not substantially changed; the basic design philosophy and implementation of the early constructors still being used today.

The smaller compact digital SHG is more appealing to the capabilities of today's amateurs and few, if any, new SHS projects are being considered. Before looking at examples of the amateur SHS we need to re-visit the methods employed by both the professional and amateur to project the solar image into the larger stationary SHG/SHS instruments. Many different solutions have been designed and used over the years, these are described below.

11.1 Cœlostat, Heliostats and Siderostats

The stationary original SHG incorporated a mirror system to direct the Sun's light into the telescope objective. The professional tower solar telescopes directed the beam vertically, whereas most of the amateur set-ups worked horizontally. The simplest method of achieving this is to use a large first surface mirror larger than the telescope objective to reflect the sunlight to the telescope. Later improvements to provide a non-rotating solar image required a two mirror solution where one was driven to compensate for the earth's rotation.

© Springer International Publishing Switzerland 2016
K.M. Harrison, *Imaging Sunlight Using a Digital Spectroheliograph*,
The Patrick Moore Practical Astronomy Series, DOI 10.1007/978-3-319-24874-5_11

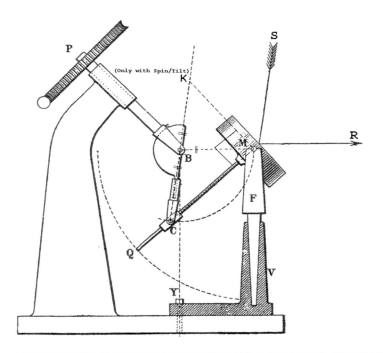

Fig. 11.1 Foucault siderostat (WIKI)

Physicist Jean Léon Foucault was the first to develop a sophisticated linkage system to allow a single mirror to be driven and present a fixed beam to a horizontal telescope as shown in Fig. 11.1. Called a siderostat, it was later supplanted by the heliostat.

Siderostats and heliostats both incorporate a flat single mirror directing the solar beam into the objective. The resulting image rotates with time and is not a suitable system of SHGs. It is however possible to incorporate a Dove prism in the optical path and drive this at half sidereal rate to compensate for the field rotation. Finding large enough prisms to avoid vignetting is a challenge.

This system can be made to work in the SHS application as the resulting image rotation can be compensated by the visual observer. For an SHS implementation the simplified version where the single front surface mirror is located in front and aligned with the optical axis of the SHS telescope works well.

Many of these Foucault siderostats were used by early solar observatories around the world, but were quickly replaced by the easier to construct and maintain cœlostat design. There doesn't appear to be an amateur implementation of the Foucault design.

Cœlostats are designed on the two mirror system to allow sky coverage without field rotation. Two flat mirrors are arranged such that the first primary mirror mounted on a polar axis directs the beam to a secondary mirror which reflects the

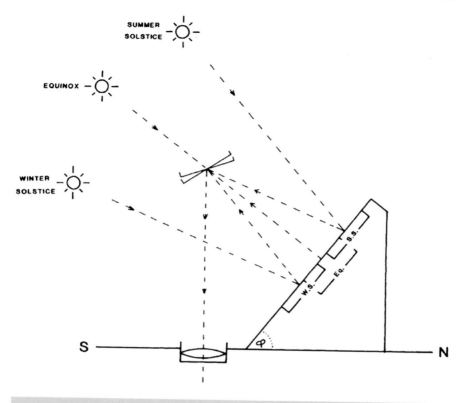

Fig. 11.2 Declination corrections in a cœlostat—movement of the primary mirror. *ws* winter Sun, *ss* Summer Sun, φ latitude (vertical telescope version shown) (A.A Mills, JBAA)

light beam to a horizontal telescope. The primary mirror rotates at half the sidereal rate, i.e. one rotation per 48 h, and must not be tilted relative to the polar axis (see Fig. 11.2).

Using a cœlostat for solar observing requires the addition of secondary movements to allow the sunlight to be presented to the telescope objective without obstruction throughout the year.

To accommodate the varying declination of the Sun (±23.5°) during the year, the position of the primary mirror must be raised/lowered along the polar axis or moved along a rail aligned with the meridian as shown in Figs. 11.2 and 11.3.

To avoid shadowing of the primary mirror by the secondary, around noon the support platform of the primary mirror must be moveable along a rail at right angles to the polar axis. The SHS examples from Slaton and Bartolick (Sects. 11.3.3 and 11.3.4) make use of the cœlostat design.

Fig. 11.3 Cœlostat—primary mirror moves along the slots shown to correct for solar declination (Bartolick)

11.2   Amateur SHS Instruments

Since the 1930s there have been many SHSs built, most of them following the Hale design. Hale's original design (see Fig. 4.6) used a mirror system to present the solar disk to the telescope, and a long focal length spectrograph fitted with both an entrance slit and an exit slit.

11.2.1   SHS Synthesizers

To visually observe an image of the Sun in the target wavelength the slit plate assembly (both entrance and exit) were mounted on an oscillating mechanism or image synthesizer. The rate of movement of about 30–40 oscillations per sec was set to give a continuous visual image of the Sun as the entrance slit moves across the solar disk The original design used a fixed solar image and moving slit plates, later designs used fixed slits and scanned the solar image. A typical early image recorded by the SHS is illustrated in Fig. 11.4.

A successful image synthesizer must meet two other key requirements. The first is the compensation for "Hα lag", which only affects vibrating slit designs. This is noted by the apparent shifting (approximately 0.4 Å) of the Hα line as viewed

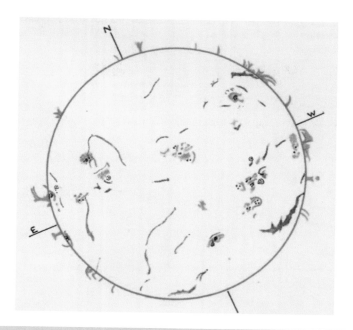

Fig. 11.4 Sketch in Hα light from an early SHS (Newbegin)

through the second exit slit. The lateral movement of the entrance slit centerline throws the slit gap away from the optical axis. Any SHS scanning arrangement that moves the slits relative to the spectrograph optics (collimator, grating, etc.) will result in a changing centreline wavelength as the slits oscillate. The origin of this effect lies in the grating equation. This equation contains the relation sinα ±sinβ, where α and β are the incident and diffracted angles and + or − depending on whether α and β are on the same side or opposite sides of grating normal. If the position of the slits move with respect to the grating then the angles α and β are changing. This means that sin α and sin β are changing, and although the changes are small the application of the grating equation will highlight the change to the incident angle of the collimated beam onto the grating and a subsequent relative movement of the Hα line to the exit slit. As the slit gaps and line widths we use are small, the deviation from the ideal can show up quickly.

The usual way of compensation for this "Hα lag' effect is to incorporate a line shifter, a thin (3–5 mm thick) plain glass plate which can be tilting to effect a shift in the central wavelength.

The original oscillating bar of the Hale design was also adapted to take advantage of the fixed slit option offered by the use of Anderson prisms. These are glass prisms approximately 15–18 mm square and 25 mm high which rotate at a fixed speed to give a full scan of the solar disk. The prisms are usually driven at 500 rpm (or greater) to give 30 (or more) solar images per second. One is positioned immediately in front of the entrance slit and another immediately behind the exit slit.

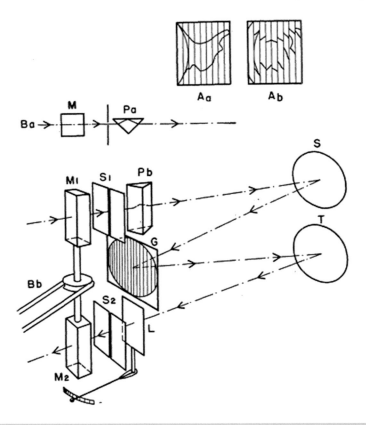

Fig. 11.5 Pettit SHS arrangement (Pettit)

The second key requirement, the correspondence condition which only affects Anderson prism designs, is shown in Fig. 11.5.

The image, Fig. 11.5 Aa shows the build-up of the slices across the image when the image synthesizer is in action. Each successive strip must match up with the previous to give a consistent final image. This can only be achieved where there are even numbers of reflections within the spectrograph. Figure 11.5 A_b shows the result when there are an odd number of reflections—each section is mirror reversed. With narrow slit gaps this is not as noticeable, but best resolution will always come from an even number of reflections. For example a Littrow arrangement has only one reflection—from the grating, an Ebert–Fastie has three—two from the main mirror and one from the grating. Pettit corrected for this condition by adding a totally internally reflective prism immediately after the entrance slit, but depending on the mechanical layout of the spectrograph a small front surface mirror would do the job.

Figure 11.5 also shows the arrangement used by Pettit, with the Anderson prisms. Bb shows the relative positions of the Anderson prisms (M1, M2), the entrance and exit slits (S1, S2) the reversing prism (Pa), the Collimator mirror (S), the grating (G), the imaging mirror (T), and the line shifter (L).

Other image synthesizer designs were introduced over the years. The Sellers vibrating fork concept was introduced in the 1930s and the HYLOV (Higgins, Young, Lovato, Ohnishi and Veio) nodding mirrors—two small mirrors mounted on a bar and driven by a small cam, was first constructed by Young in 1992.

These options were used by various amateurs around the world as they enjoyed for the first time the visual aspects of solar Hα observing. Fred Veio's self published book "The Spectrohelioscope", first issued in 1972, became the bible for aspiring amateurs interested in the SHS and it is still recommended reading today. A free eBook copy of the latest version is available courtesy of Veio on the Yahoo group Astronomical Spectroscopy website (see Appendix D).

Unfortunately, due to the size and complexity of the SHS it has fallen out of favor with the DIY amateurs and the advent of the compact, digital SHG has almost made the original SHS redundant.

11.2.2 Veio

Fred Veio's name is synonymous with amateur SHS. For the past 50 years Fred has been a continuing source of information and support to the many amateurs taking up the technical challenge of building a SHS. His book on the design and construction of the SHS has become the de-facto design manual for many novices.

Veio has constructed and modified many SHSs over the years. Generally they have followed the basic Hale design but used a Heliostat feeder, long focal length lenses, and high dispersion gratings. Veio initially tried the Hale vibrating slit assembly, but then developed his own version of the rotating slit disk synthesizer (see Fig. 11.6). The original rotating disk invented by Stanley in 1912 made use of 150 slots and was 200 mm diameter. It was later trialed by Hale (using 50 slits) for use on the Mt. Wilson 60 ft tower telescope but he did not fully pursue the application.

The metal disk has 24 by 200 μm wide slots and acts as a shutter allowing either the slots in the glass disk to be exposed or the focusing/alignment slots (shown dotted in Fig. 11.6). Initially the metal shutter is rotated and the single 10 μm slit used to produce a spectrum in the eyepiece. The target wavelength can then be viewed and brought to the center of the exit window by tilting the grating. The metal shutter is then rotated to expose the glass slits. The 24 slots on the 110 mm diameter glass disk are 125 μm wide and the disk rotates at 60 rpm (one rev per second) to give persistence of vision.

Veio's early SHS was well described in a Sky and Telescope, January 1969 article (Fig. 11.7). It utilized a Heliostat (Fig. 11.8) and had a first surface Pyrex mirror, 1/15 wave, 115 mm×68.5 mm. This fed a single lens objective with 63 mm

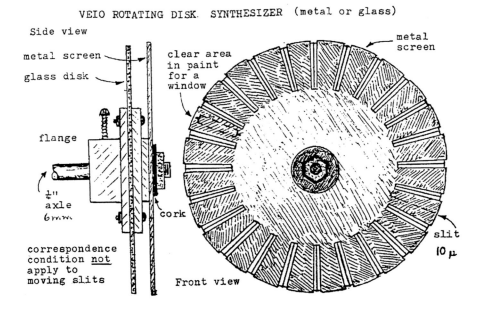

Fig. 11.6 Veio combination glass slit disk (Veio)

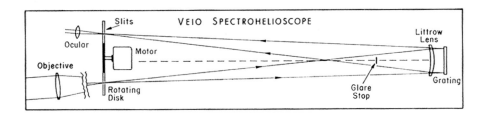

Fig. 11.7 General arrangement of the SHS (Veio)

diameter, 2781 mm focal length. The slit disk assembly, driven by a small 7 W synchronous motor at 60 rpm, provided an ideal rate for persistence of vision. Another single lens was used for the Littrow collimator, 50 mm diameter, 1905 mm focal length. The grating used was a Bosch and Lomb 32 mm × 30 mm, 1200 l/mm, blazed for 5500 Å. This gave a linear dispersion of 4 Å/mm.

For visual observing, a long focal length (115 mm fl.) eyepiece was constructed. This gave a magnification of ×24 and provided excellent views of the solar surface in Hα.

Veio always encouraged amateurs to build the SHS cheaply. Buy the best grating and quality optics then spend time rather than money on the actual SHS construction. He said his first SHS cost only $265 (1969). This was still active in 2003 as shown in Fig. 11.9.

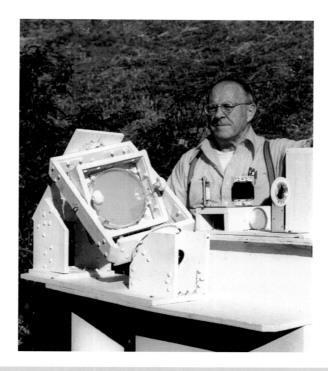

Fig. 11.8 Fred Veio at the heliostat of the SHS (Veio)

Fig. 11.9 Veio SHS circa 2003 (Veio)

Veio and Higgins were the first recorded amateurs to use an SHS to replicate Hale's early work on solar magnetism. In 1999 they successfully recorded the Zeeman Effect.

11.2.3   Slaton SHS

Like many other amateurs, Jon Slaton's introduction to the SHS was through Fred Veio's book. In 1998 he decided to build a Hale type SHS based on a 6″ (150 mm), 7500 mm focal length, f50 lens (parfocal at C and K wavelengths) he found for sale online. Around the same time he purchased the mechanicals of a large Gaertner cœlostat. This turned out to be a Hale type cœlostat, designed for a 10″ (250 mm) primary mirror and an 8.75″ (222 mm) secondary (see Fig. 11.10). The missing mirrors were replaced and further parts acquired; a pair of matched 6″ (150 mm), 4125 mm focal length (f27.5) mirrors, a 3″ (75 mm) diagonal some Anderson prisms and a 4″ (100 mm) square 600 line grating engraved on a 6″ (150 mm) diameter flat. These items then all came together for the construction of his SHS.

Fig. 11.10 Hale cœlostat (Slaton)

To accommodate the instrument in the available space an 11″ (280 mm) first surface flat mirror was added to direct the cœlostat beam towards the telescope and his building. In the early testing phase it became obvious that the cœlostat and other optical elements would have to be fitted with motors. A series of motors were installed, wired back to a central controller at the eyepiece. The primary cœlostat mirror was motorized using an 18″ (457 mm) worm drive and the secondary mirror with two motors to give North/South and East/West motions.

The added 280 mm flat mirror was also fitted with two motors—one for up/down, the other for left/right to direct the solar image into the telescope objective, which is also fitted with a focusing motor. The long focal length of the telescope lens produces a solar image of about 2.75″ (70 mm). Just before the entrance slit, the light goes through the Anderson prism assembly, which was mounted as close as possible to the slit plate.

The Anderson prisms are mounted on shaft driven by a variable speed motor (Fig. 11.11). A dovetail system was incorporated to allow the rotating prisms to be adjusted or moved as necessary (Fig. 11.12). The prisms have to be moved out of the way to adjust the slits; this is a simple operation only taking a minute or so.

Making the slit plates proved to be a challenge. The initial trials using the slit design by John Strong (described in ATM, Book III, p. 144) proved difficult to construct, and although it worked nicely, turned out to be too big physically for the available space. In the end 4″ (100 mm) PVC inspection caps were used as a support for strips of refrigerator magnet material. The magnetic strips were glued on either side of a 3″ × ¼″ (75 mm × 6 mm) long central slit gap cut in the caps These strips hold securely the 4″ (100 mm) wallpaper scraper blades in place on the side of the cap facing the rotating prisms. The center line is clearly marked. The PVC

Fig. 11.11 Anderson prism assemblies (Slaton)

Fig. 11.12 Anderson prism assemblies in position at slit plates (Slaton)

Fig. 11.13 Entrance and exit slit plates (Slaton)

caps shown in Fig. 11.13 proved to be easy to remove, to adjust the slit width, and also easily rotated to achieve good slit alignment.

The collimating and imaging mirrors are mounted on a fully adjustable track system. These matched mirrors sit on a common platform with two motors—one to

Fig. 11.14 Spectrograph collimating and imaging mirrors on focusing carriage (Slaton)

move both mirrors at the same time; and one to move one mirror independently for fine focus. These are shown in Fig. 11.14.

The collimating mirror is focused on the back side of the first slit and reflects the parallel beam to the 100 mm grating which sits just above the slits.

The grating platform (Fig. 11.15) is fitted with micrometer controls. The grating has 3 micrometer controls, along with a motor for fine focus. It is also adjustable for rotation. The imaging mirror takes the dispersed light from the grating, and brings it to a focus on the back of the second slit. A long triangular prism to invert the light was added just before the second slit. This is required to correct the correspondence condition which only affects Anderson prism designs.

The second Anderson rotating prism is positioned on the front side of this slit, then a 3″ (75 mm) diagonal, and finally the eyepiece which is focused on the slit. This arrangement is shown in Fig. 11.16. The action of the second rotating Anderson prism provides the synthesized image.

The motors used for guiding, focus, and speed of the rotating prisms can all be controlled while standing at the eyepiece position. This makes the operation of the SHS very easy. The 280 mm flat is used to scan the perimeter of the Sun.

To control the tracking precisely, Slaton made a ring with 4 light sensitive diodes oriented North, South, East, and West. This ring sits just over the 280 mm flat. When one diode senses the Sunlight, it operates a motor relay to slowly move in one direction on the secondary tower. As soon as the light leaves the diode, it shuts off the motor. Should it go too far, the opposite diode slowly operates the same motor in the reverse direction. The other two diodes work the same way, but in the other two

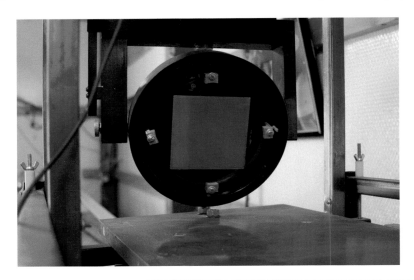

Fig. 11.15 Grating holder assembly (Slaton)

Fig. 11.16 Final assembly of slit/ prisms and visual diagonal (Slaton)

directions with the second motor. The diodes are repositioned at the extreme edge of the light path from the secondary mirror at the start of each observing session. The slits are only opened a few microns or so, allowing a very tiny slice of the Sun to be seen at one time were it not for the rotating prisms. Where the light beam enters the building a 6″ stovepipe painted flat black on the inside was added. The prisms are located inside it and a removable smaller section covering the prisms allows for maintenance.

11.2.4   Higgins SHS

Leonard Higgins built his SHS again following the basic designs of Veio. In 1989 a new heliostat was incorporated using a 150 mm diameter ¼ wave front surface flat. It is fitted with a right ascension drive (running at half sidereal rate, see Sect. 11.3) and a declination drive to direct the image of the Sun into the long focus reflecting telescope. Both these drives are controlled by a remote control paddle.

A spherical 125 mm diameter, 2740 mm focal length mirror is used in conjunction with a 50 mm, 500 mm negative focal length Barlow lens to give an effective focal length of 6000 mm with a solar image of 54 mm diameter (see Fig. 11.17).

Fig. 11.17 General arrangement of the SHS (Higgins)

Fig. 11.18 Details of the entrance and exit slit plates and the nodding mirror assembly (Higgins)

The beam is then folded by a 45° front surface mirror before reaching the nodding mirror assembly which sits immediately in front of the entrance slit.

Figure 11.18 (left side) shows the entrance and exit slit arrangement and the (right side) nodding mirror mechanism. The adjustable slits were constructed to Barnes and Brattain's design (see ATM, book II, p. 503).

The nodding mirror image synthesizer is the heart of the Higgins SHS. It is made up from two front surface mirrors mounted on a lightweight frame which is then oscillated in front of the entrance slit and behind the exit slit to scan the solar image across the slit gap. This assembly needs to be made so it will induce little or no vibration in the optical light path of the spectroscope section which could create movement of the air currents, degrading the quality of the image. A heavy steel plate was used as the mounting for the small electric motor with eccentric wheel— cam and the vertical supports for the aluminum nodding plate that holds the two mirrors. The steel plate is then spring mounted on a large aluminum base which is attached to the wooden box of the SHS and cushioned with foam. The idea is to isolate as much vibration as possible. The follower that rides on the eccentric wheel and oscillates the mirror plate was made from a discarded set of automobile ignition points. The fiber section was used and a small amount of cam lubricant is applied

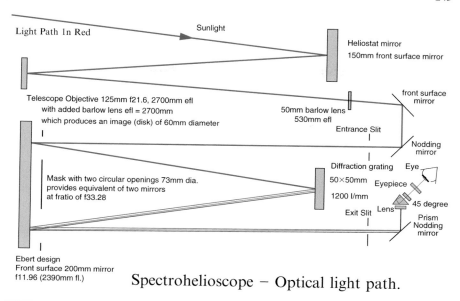

Spectrohelioscope – Optical light path.

Fig. 11.19 SHS Optical layout (after Higgins)

on occasions. The distance between the nodding mirrors and the slits is 150 mm. The nodding mirror system works well for Higgins.

The spectrograph is an Ebert–Fastie design using a 200 mm f11.96 spherical mirror, focal length 2430 mm. A baffle disk with two apertures, 73 mm diameter is positioned at the main mirror giving the collimating and imaging beam an effective f33.28 (see Fig. 11.19).

A 50 mm×50 mm 1200 l/mm blazed grating is used (in the first or second order) and the grating support is fitted with a worm gear system capable of rotating the grating from the first order through to the third.

The diffracted beam then goes through the exit slit gap where it is reflected by the nodding mirror to the eyepiece assembly. This uses a 45° erecting prism mounted between two small achromatic lens of 89 mm focal length, and when combined with a 26 mm Plossl eyepiece. The final arrangement performs very well and gives good visual results.

Results

Higgins' sketch of the solar disk in Hα made on June 27 1999 (Fig. 11.20), shows the typical visual detail which can be observed. The filaments and prominences are well defined.

11.2.5   Bartolick

Although fully described in Sect. 9.11 as a digital SHG, Bartolick's original SHS layout should be acknowledged and recognized as a prime example of the classic SHS design. In his SHS configuration, the key elements are all there—the cœlostat, nodding mirrors, entrance/exit slit assemblies, line shifter and even the smile correction lens.

To provide the image synthesizer in the SHS Bartolick added a vibrating, or nodding, mirror arrangement, driven by a General Scanning G330 galvanometer scanner (Fig. 11.21).

As an alternative a similar drive system was also trialed with a vibrating slit plate. The base was about 100 mm long that had much of its mass machined off, with small stiffening ribs remaining. At each end there was a 12 mm long slit with jaws made from 0.010 in. thick steel feeler gauge stock. A hole in the center fitted the shaft of the galvanometer scanner. Getting the 30 μm slits the same width and lined up was a challenge. To set the slit positions, the bar was held in the vice of a milling machine and a microscope mounted in the spindle. By moving the table, the jaw alignment could be inspected and adjustments carried out.

All moving slit systems have problems with scanning in a fixed wavelength as the slits moves relative to the optical axis when they oscillate back and forth. This

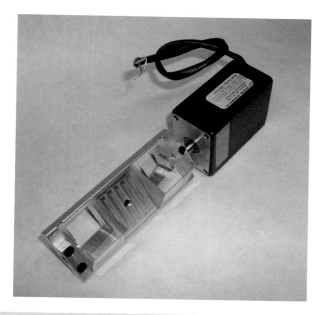

Fig. 11.21 Double prism oscillating plate (nodding mirrors) used in the SHS (Bartolick)

causes the "Hα lag" discussed earlier. Unfortunately the end result was a disappointment and did not live up to expectations, and the design was returned to the nodding mirror synthesizer arrangement.

Bartolick also makes the comment that all SHS by design tend to have large f numbers, and with large f numbers come large Airy disks. A good rule of thumb is that in the middle of the visual range (5000 Å) the Airy disk diameter in μm of an imaging system is approximately equal to the F number (focal ratio) of the system. This grows to approximately 1.2× the f number at Hα.

The final resolution and overall performance of the instrument PSF will reflect the larger Airy disks and is the reason most SHS/SHG builders tend to overestimate the narrowness of their bandpasses. Whatever the approach, the spectrohelioscope is an interesting if under-explored area in DIY amateur astronomy.

Further Reading

Sidgwick. J.B.: Amateur Astronomers Handbook, Faber and Faber (1971)
Ingalls, A.G., (Ed): Amateur Telescope Making Vol1. Scientific American (1980)
Ingalls, A.G., (Ed): Amateur Telescope Making Vol2. Scientific American (1978)
Sky and Telescope, **34**, p329 (1967)
Ibid, **39**, p215 (1970)
Mills, A.A.: "Heliostats, Siderostats, and Cœlostats: A review of practical instruments for Astronomical applications", JBAA, 85, p 89 (1985)

Webpages

http://articles.adsabs.harvard.edu//full/1939PASP...51...95P/0000096.000.html
http://www.eyes-on-the-skies.org/shs/
http://www.astrosurf.com/cieldelabrie/sphelio.htm (In French)
http://www.astronomie-amateur.fr/Projets%20Spectro1.html (In French)

Chapter 12

Future Solar Imaging Developments

12.1   Solar Imaging

Amateurs will always continue to image the Sun in white light. The safety solar film and Herschel wedge options are here to stay. These filters provide a highly capable, cost effective and a safe environment for basic solar imaging. Future improvements will probably not come from the filter technology but from the use of the next generation of fast-frame cameras. These cameras have the potential to record added high resolution detail to sunspot images, surrounding fibrils, and faculae and give amateurs a very satisfying result. Before too long the solar imager will however look towards new challenges.

It's fair to say that the commercial introduction of the Hα etalon solar telescopes and filters dramatically changed amateur solar imaging taking them from the decades of old white light sunspot images to the exciting new world of narrowband imaging of the chromosphere. Add the new digital and the fast-frame cameras to the mix and you have one of the most rapidly growing markets in amateur astronomy. A survey of popular astronomical forums around the world would show more and more solar images being presented.

The key players like Coronado and Lunt followed by DayStar with their Quark filter continue to dominate the amateur market. It's been over 15 years since the early Coronado Hα filters appeared and other than the introduction of the Richview and Lunt's pressure tuning refinements not much has changed. The technology is mature and the cost of entry to any new company trying to gain a foothold in this lucrative niche market is therefore very high.

One common area of concern with all current etalon Hα designs is the high failure rate of the ITF filter used in the blocking filter assembly. It is not unusual to

© Springer International Publishing Switzerland 2016
K.M. Harrison, *Imaging Sunlight Using a Digital Spectroheliograph*,
The Patrick Moore Practical Astronomy Series, DOI 10.1007/978-3-319-24874-5_12

hear of these ITF filters deteriorating within months and they usually appear to have a limited lifespan of 3–5 years. This fact at least has been recognised by Daystar who offers, at a cost, a return to base replacement. It is very disconcerting for amateurs to find that solar telescopes costing thousands of dollars can be subjected to this random failure. This is certainly one area of product improvement which would be welcomed by the solar community.

A limitation of the air-spaced etalon filter as used by Coronado and Lunt is the inability to accurately determine the actual central wavelength during the tuning of the filter. It would add a very useful scientific dimension to the images produced to be able to accurately and with repeatability position the target wavelength into known positions within the red/blue wings. This is currently achieved on the solid etalon DayStar Quantum filter by using a temperature controlled oven.

Even with its current limitations, the Hα solar filter continues to gain popularity with the amateur. The latest generation of fast frame cameras with USB3 connectivity provide exciting opportunities of combining larger frame sizes with faster frame rates. This combination gives the potential to further improve the quality of the images produced and reducing adverse seeing effects. It is supported by the ongoing software development of programs like Registax and Autosterkkert, meaning we will continue to see solar images demonstrating even higher spatial resolution.

An interesting area of endeavor now becoming more popular with amateurs is the presentation of time lapse video movies. We see the professional solar movies from NASA and Big Bear Solar Observatory among others appearing more and more often in science-oriented Web presentations and on TV; it's now become possible for the amateur to do something similar with the equipment they already have. A series of Hα exposures taken every 5 min over a 2 h period and processed by a basic movie maker software like Virtual Dub, or Windows Movie Maker can result in a very spectacular and thought provoking movie. The dynamics of the magnetic fields around sunspots and the growing and blossoming of the prominences becomes even more impressive viewed in this way. Over the next few years expect to be amazed and enthralled by the amateur movies of the Sun in Hα.

12.2   Amateur SHG Developments

The spectroheliograph has been in existence now for over 120 years, almost as long as astronomical photography, and its contribution to solar observing is nothing less than revolutionary. The early discovery of the solar magnetic field through the observation and measurements of the Zeeman effect, the bipolar nature of sunspots and the subsequent 22-year magnetic cycle changed forever our ideas of the construction and dynamics of the Sun. The spectroheliograms obtained at the beginning of the twentieth century gave insight into the chromosphere and showed a wealth of features never imagined by previous generations. Until the 1940s the spectroheliograph used on the large solar tower telescopes around the world dominated the leading edge technology for solar studies. It was the advent of specialized narrowband filters like the Lyot filter and the Fabry–Perot etalons which gradually pushed the spectroheliograph to the background. Large spectrographs are still used to record the solar spectrum but the magic of the spectroheliograph has gone.

It is only through the work of Fred Veio and the small but dedicated group of spectrohelioscope builders that we are able to relive and enjoy the benefits of the spectroheliograph today. But for Fred's continued passion to support amateurs and spread his knowledge and his practical experience of building and using the visual spectrohelioscope we would not be in the comfortable position we find ourselves in today. The concept and examples of the spectroheliograph would be a museum relic otherwise—a collection of glass and brass gathering dust behind a fading signboard. The revitalized spectroheliograph has been given new life in the digital era and is once again being recognised as a very versatile narrowband imaging solution to a new generation of amateur solar observers.

We also owe the inventor of the humble webcam a tremendous debt of gratitude. The introduction of the small 640×480 pixel array camera and frame rates of up to 10 fps opened up imaging opportunities to many amateur astronomers. Many amateurs will remember their very first digital image of the moon or Jupiter. Consumer demand has pushed the ongoing development of the webcam and new generation imaging sensors with 3 or 4 million pixel arrays can be found in the average digital mobile phone. Industry has also embraced the use of fast frame imaging systems for routine inspection and quality control on manufacturing and assembly lines. Fortunately for the amateur, these developments have resulted in the availability of quality cameras which can be easily adapted to astronomy, and our specialised area of solar imaging. If we were the only market for such innovations the costs of utilizing any newly developed camera technology would be beyond our means, and our wallets.

The construction of a digital spectroheliograph, based on readily available commercial components (lenses, gratings etc.) is not a difficult task and well within the capabilities of the average handyman. There's a sense of satisfaction when the first spectrogram appears on the computer screen. The challenges of being able to actually determine the resolution and bandwidth being achieved come later and the excitement of being able to preparing images in other significant wavelengths just by rotating the grating within the spectrograph is exciting.

As the use of the digital SHG gains momentum and new instruments come into use the boundaries keep shifting. The early digital SHG were conversions from the old large SHS designs. Then, as the benefits of the small pixel CCD camera were recognised, the telescope and spectrograph focal lengths could be reduced, resulting in smaller more flexible designs which were transportable and could be mounted on equatorial mounts. The latest digital SHG as typified by Zetner and Smith show that even smaller apertures and focal lengths can be combined with fast scanning rates to produce excellent images. As the fast-frame and ROI capture capabilities of the CCD cameras continue to improve so does the future opportunities. The potential to obtain real time imaging is almost within our grasp and the improved quality being obtained with the current processing software is exciting.

Many of the amateur SHG software developers continue to work closely with the small but dedicated user group and bring out regular refinements to further enhance the processing capabilities. It would not be surprised to see more and more solar observers taking up the challenge of the digital SHG and going somewhere towards replacing the use of commercial narrowband filters within the next few years.

12.3 Commercial SHG

There is at least one version of the digital SHG being considered for commercial production. Based on Mete's hi-res Littrow design and developed by Avalon Instruments —the Solarscan. Figure 12.1. This will utilize an 80 mm, 400 mm focal length refractor, an adjustable slit and the Littrow spectrograph based on a 600 mm focal length refractor and a 2400 l/mm grating. The assembly is expected be about 1150 mm long and weigh 12 kg.

Fig. 12.1 The Solarscan prototype on an Avalon M1 equatorial mount (Mete)

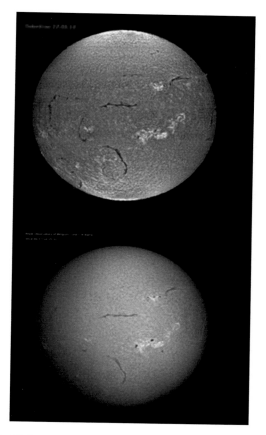

Fig. 12.2 Hα disk using the SolarScan prototype 27 Sept. 2014 (Mete)

This instrument will incorporate unique electronic features and a new software package interfacing with the Avalon Star Go. Features include the automated control of the entrance slit aperture to a precision of 10 μm and the inclusion of high resolution motorized focus on both optics, of the telescope and spectrograph. The software being developed, when fully implemented, is expected to be able to be programmed to control these features as a function of wavelength and the camera sensor being used. It will also provide automatic search and positioning on the spectral line of interest. It is anticipated that further software will be developed to handle the processing of the solar image extracted from the selected line of interest.

Figure 12.2 shows the comparison of a prototype Hα image to that obtained at the professional solar Observatory, The Royal Observatory of Belgium.

With all of these advances, the opportunities for increased amateur astronomy focused on the Sun have a very bright future.

Further Reading

Karttunen, H. et al (Ed.): Fundamental Astronomy, 4th Edition. p214–p216. Springer (2003)

Webpages

https://archive.org/details/BaaMemoirsVol37Part2-TheSun
http://www.pno-astronomy.com/
http://www.avalon-instruments.com/
http://fiss.snu.ac.kr/db/papers/files/FISS_chae.pdf
http://sos.kasi.re.kr/english/instrument/ksis.php
http://www.bbso.njit.edu/newtelescope/large.html

Appendix A

Summary of Important Solar Wavelengths

A.1 Fraunhofer Lines

Name	Wavelength (Å)	Source
A-band	7594–7621	Atmospheric O_2
B-band	6867–6884	Atmospheric O_2
C	6563	$H\alpha$
a-band	6276–6287	Atmospheric O_2
D1	5896	Na I
D2	5890	Na I
E	5269	Fe I
b1	5184	Mg I
b2	5173	Mg I
c	4958	Fe I
F	4861	$H\beta$
d	4668	Fe I
e	4383	Fe I
f	4340	$H\gamma$
G	4308	Ca/Fe (430.77/430.79)
g	4227	Ca I
h	4102	$H\delta$
H	3968	Ca II
K	3934	Ca II

© Springer International Publishing Switzerland 2016
K.M. Harrison, *Imaging Sunlight Using a Digital Spectroheliograph*,
The Patrick Moore Practical Astronomy Series, DOI 10.1007/978-3-319-24874-5

A.2 *Major Solar Spectral Lines*

Wavelength (Å)	Name	Wavelength (Å)	Name
3928	Fe I	4271	Fe I
3934	**CaII (CaK)**	4299	Fe I
3944	Al I	4308	Fe I
3951	Fe I	4323	CH
3953	Fe I	4324	CH
3957	Fe I	**4340**	**Hγ**
3961	Al I	4469	Ti
3968	**CaII (CaH)**	4554	Ba II
3970	**Hε**	4665	Fe I
3878	Fe I	4782	Mn
3987	Mg I	**4861**	**Hβ**
3997	Fe I	4932	Ba
4005	Fe I	4956	Fe I
4031	Mn I	5036	Ni
4033	Mn I	**5167**	**Mg** (Magnesium triplet)
4034	Mn I	**5173**	**Mg**
4036	Mn I	**5184**	**Mg**
4046	Fe I	5270	Fe I
4055	Mn I	5328	Fe I
4057	Mg I	5455	Fe I
4064	Fe I	5456	Fe I
4102	**Hδ**	5528	Mg
4119	Ca I	5614	Fe I
4132	Fe I	5707	Fe I
4144	Fe I	5709	Fe I/Ni I
4154	Fe I	5711	Mg
4167	Mg	5786	Cr
4173	Fe I	**5873**	**He (D3)**
4191	Fe I	**5890**	**NaI** (Sodium doublet)
4198	Fe I	**5896**	**Na I**
4202	Fe I	5948	Si I
4215	Sr II	6180	Ni
4227	Ca I	6279	O_2 (Atmos.)
4236	Fe I	6358	Zn
4247	Sc II	6398	Fe I
4254	Cr I	6497	Ba II
4260	Fe I	**6563**	**Hα**
		6847–6944	O_2 (Atmos)

A.3 Suitable Target Lines for Zeeman Effect

Wavelength (Å)	Name	Landé factor
8468.417	Fe	2.49
6767.785	Fe	1.43
6302.508	Fe	2.49
6301.505	Fe	1.5
6258.705	Fe	1.11
6173.348	Fe	2.5
6102.735	Ca	2.0
5324.184	Fe	1.67
5250.218	Fe	3.0
5247.576	Cr	2.51
5131.478	Fe	2.5

Appendix B

SimSpec and SimSpec SHG Design Spreadsheets

The spectroscope design spreadsheet, SimSpec has been widely used over the past few years by amateurs designing, or validating their spectroscope design concepts. It provides a very comprehensive and accurate assessment of instrument performance. A version of the SimSpec spectrograph design spreadsheet, SimSpec SHG has been prepared for SHG applications. This will allow you to calculate the sizes and layout of the components in your proposed design and the probable outcomes. Estimates of anamorphic factor, dispersion and resolution are also given.

These spreadsheets are available for download in the files area of the Astronomical Spectroscopy Yahoo Group.

© Springer International Publishing Switzerland 2016
K.M. Harrison, *Imaging Sunlight Using a Digital Spectroheliograph*,
The Patrick Moore Practical Astronomy Series, DOI 10.1007/978-3-319-24874-5

Appendix C

Suppliers of Components of Interest to the Amateur

JTW Spectra-L200

JTW Astronomy
Aalsmeerderweg 103M, 1432CJ Aalsmeer, The Netherlands
http://www.jtwastronomy.com/products/spectroscopymain.html

Shelyak Lhires III

Shelyak Instruments
Les Roussets, 38420 Revel, France
http://www.shelyak.com/

Gratings

Optometrics.
8 Nemco Way, Stony Brook Industrial Park, Ayer, MA 01432, USA
   http://www.optometrics.com/

Thorlabs, Inc.
Thorlabs—Newton, New Jersey
435 Route 206 Northm, Newton, NJ 07860, USA
   http://www.thorlabs.de/

Newport Corp. (Richardson Gratings)
705 St. Paul St, Rochester, NY 14605, USA
http://www.newport.com/products/overview.aspx?sec=124&lang=1033

© Springer International Publishing Switzerland 2016
K.M. Harrison, *Imaging Sunlight Using a Digital Spectroheliograph,*
The Patrick Moore Practical Astronomy Series, DOI 10.1007/978-3-319-24874-5

Prisms

Optometrics (see above)
Edmund Optics, Inc.
101 East Gloucester Pike, Barrington, NJ 08007-1380, USA
http://www.edmundoptics.com/
Surplus Shed
1050 Maidencreek Road, Fleetwood, PA 19522, USA
http://www.surplusshed.com/

Entrance Slits

Surplus Shed (see above)
JTW Astronomy (see above)
Shelyak Instruments (see above)
Thorlabs (see above)
Lennox Laser
12530 Manor Road, Glen Arm, MD 21057, USA
http://www.lenoxlaser.com/

Lenses and Mirrors

Knight Optics
Unit 4, Roebuck Business Park, Harrietsham, Kent ME17 1AB, UK
http://www.knightoptical.com/
Surplus Shed (see above)
Edmund Optics (see above)
Thorlabs (see above)

Appendix D

**Useful Solar/
Spectroscopy Forums**

https://groups.yahoo.com/neo/groups/astronomical_spectroscopy/info
https://groups.yahoo.com/neo/groups/spectrohelioscopes/conversations/messages
https://groups.yahoo.com/neo/groups/jtwspectroscopy/conversations/messages
http://solarchat.natca.net/ (Charlie Bates Solar Astronomy Project)
https://groups.yahoo.com/neo/groups/Iris_software/info
https://groups.yahoo.com/neo/groups/spectro-l/info
http://spektroskopie.fg-vds.de/
http://www.astrospectroscopy.de/

Other Websites

http://www.astrosurf.com/cieldelabrie/sphelio.htm (In French)
http://www.astrosurf.com/rondi/obs/shg/index.htm
http://www.astrosurf.com/spectrohelio/index-en.php
http://www.astrosurf.com/joseribeiro/Eespectroheliografia.htm
http://www.lightfrominfinity.org/Hirss2%20spettroelioscopio/hirss2_spettroe-
lioscopio.htm
http://www.astrosurf.com/buil/lhires2_Sun/first.htm
http://astrosurf.com/buil/us/book2.htm
http://www.ursusmajor.ch/astrospektroskopie/richard-walkers-page/index.html

© Springer International Publishing Switzerland 2016
K.M. Harrison, *Imaging Sunlight Using a Digital Spectroheliograph*,
The Patrick Moore Practical Astronomy Series, DOI 10.1007/978-3-319-24874-5

Appendix E

Selected Bibliography

Becker, B.J.: Unravelling Starlight, CUP (2011)
Tonkin, S.F. (Ed): Practical Amateur Spectroscopy, Springer (2002)
Robinson K.: Starlight, Springer (2009)
Howell, S.B.: Handbook of CCD Astronomy, CUP (2010)
Jackson, M.W.: Spectrum of belief, MIT Press (2000)
Thorne, A., Litzen, U. and. Johansson, S.: Spectrophysics, Springer (1999)
Scott Ireland, R.: Photoshop Astronomy. Willmann Bell (2005)
Wodaski, Ron: The New CCD Astronomy. New Astronomy Press (2002)
Covington, M.A.: Astrophotography for the Amateur. CUP (1985)
Clark, S.: The Sun Kings, Princeton University Press (2007)
Sidgwick, J.B.: Observational Astronomy for Amateurs, Faber (1957)
Martinez, P. (Ed): The Observer's Guide to Astronomy, Volume 1, CUP (1994)
McGucken, W.: Nineteenth-Century Spectroscopy, The John Hopkins Press (1969)
Maran, S.P. (Ed): The Astronomy and Astrophysics Encyclopedia, CUP (1992)
The following references/books are also available on the web for download at http://www.archive.org/. They give a wonderful insight into the early years of spectroscopic development and solar observing.
"The Spectroscope and its work"—Richard A Proctor, 1888
"The Spectroscope and its applications"—J Norman Lockyer, 1873
"How to work with the Spectroscope"—John Browning, 1882
"Sir William Huggins and Spectroscopic Astronomy"—E W Maunder, 1913

© Springer International Publishing Switzerland 2016
K.M. Harrison, *Imaging Sunlight Using a Digital Spectroheliograph*,
The Patrick Moore Practical Astronomy Series, DOI 10.1007/978-3-319-24874-5

Yahoo Technical Support Group

A support group dedicated to assisting amateurs in astronomical spectroscopy and digital SHG's has been set up by the author.

https://groups.yahoo.com/neo/groups/astronomical_spectroscopy/info

http://www.astronomicalspectroscopy.com/index.html

Membership of this group is open to any interested amateur and provides an ideal forum in which to raise questions on any and all aspects of spectroscopy/solar observing. There are many files available in the "knowledge base" covering spectroscope design and construction as well as the practical aspects of spectral imaging and subsequent analysis. Spectrum images are regularly uploaded to assist the novice and there are many experienced amateurs who are only to willing to discuss any problem you may have as you move up the learning curve.

The associated Webpage has been developed to further assist the members and will be updated regularly with information and relevant links.

Any corrections and or amendments to this book will be published on the Astronomical Spectroscopy forum. See you there …

Glossary

Absorption lines Dark lines seen in a solar spectrum. Caused by the change in energy level of an atom.

Achromatic Color correction of a lens, where at least two different wavelengths are brought to a common focus.

Active region Regions of the solar atmosphere where magnetic activity causes emission activity. Associated with the formation of sunspots and flares.

ADU Analog to Digital Unit. The measure of intensity used by CCD cameras. Related to bit depth of the chip.

Anamorphic factor Distortion of the diffracted image caused by a grating when the incident light is at an angle to the normal.

Anderson prism Square section prisms used in a pair as image synthesizers in a spectrohelioscope.

Angstrom Measure of wavelength. 1 Å is 10^{-10} m and 10 Å equals a nanometre (nm).

Angular measure In astronomy: Full circle = 24 h or 360°, 1° = 60 arc min, 1 arc min = 60 arc seconds One min of time equals 15 arc min.

APO, Apochromatic Where a lens is color corrected to bring three different wavelengths to a common focus.

Arc second/min Seconds of arc etc. See angular measure.

Aurora Display of colored light, usually visible from high northern (aurora borealis) or southern (aurora australis) regions. Caused by the interaction of the solar wind and the Earth's magnetic field.

Atmospheric Lines See Telluric lines.

Balmer series A series of absorption lines see in a stellar spectrum. Caused by the various energy levels of the element hydrogen. The strongest line (Hα) is in the red portion of the spectrum.

© Springer International Publishing Switzerland 2016
K.M. Harrison, *Imaging Sunlight Using a Digital Spectroheliograph*,
The Patrick Moore Practical Astronomy Series, DOI 10.1007/978-3-319-24874-5

Bandwidth   The measure of transmission width of a filter, in wavelengths.

Bayer matrix   A method of applying filters, normally red, green and blue to a CCD chip to allow it to record colored images.

Bias   Sometimes called read noise. The added noise signal applied to an image by the electronics controlling a CCD when the signal is transferred from the chip.

Birefringent material   A polarizing crystal like Mica or Iceland Spar which transmits a double image due to double refraction whereby a ray of light, is split by polarization into two rays taking slightly different paths.

Bit depth   The maximum intensity recorded by the pixel in a CCD. A bit depth of 16 gives a maximum ADU count of 65,000 (2^{16}).

Black Body Curve   The intensity curve of a continuum spectrum which reflects the temperature of a star.

Blazed gratings   A diffraction grating which has the shape of the grooves angled to improve the illumination of the first order spectrum.

Blocking filter   A narrowband filter applied to an etalon to suppress unwanted wavelengths.

BMP   A bit mapped image format. Commonly used in Windows applications.

Bohr magnetron   Units of measure used in magnetic field calculations. Bohr magnetron (μB). $\mu B = 0.467$ cm^{-1} T^{-1}.

Butterfly diagram   The distribution in latitude of the sunspots over the sunspot cycle.

Carpenter prism   See Grism.

CCD   Charge coupled device. Used in cameras to record photons and convert them into electrons which are then used to generate an image.

Chromatic aberration   An optical aberration in a telescope or lens where different colors come to a different focal point. See Achromatic.

Chromatic coma   See Spectral coma.

Chromosphere   The outer atmosphere of the Sun. Gives rise to absorption lines in the spectrum.

Coelostat   Two flat mirrors used to present a non-rotating view of the Sun. Usually used in large spectroheliographs.

Continuum   The continuous underlying spectrum of the Sun.

Convection zone   Region of the Sun's interior between the Radiative zone and the photosphere, where energy is transported by convection. Magnetic fields originate in this zone.

Corona   A high temperature region of ionized plasma above the chromosphere.

Coronagraph   A telescope which uses an occulting disk to block the photosphere and allow the outer solar regions to be observed.

Coronal mass ejection   Also CME. Rapid outward movement of coronal material associated with flares or erupting prominences.

Coronal rain   Apparent droplets of material failing from below large prominences. Caused by the collapse of the local magnetic field removing the support mechanism thereby allowing material to return to the solar surface.

CWL   Central wavelength of a filter transmission curve.

Cylindrical lens   A lens which only focuses in one axis. Produces a line image from a point source.

Deviation angle   The angle between the incident light ray and the emerging colored ray from a prism or grating. Varies with the wavelength of light.

Diagonal   A prism or mirror used to divert the light path in a telescope through 90°.

Dichroic filters   A multi coated filter where the coating materials cause interference and only allow a narrow bandwidth to be transmitted.

Dispersion   The ability to break white light into various colors. Normally associated with prisms and gratings.

Doppler shift   Change in wavelength frequency due to the speed of recession or approach. Also called line of sight velocity change in solar observations.

Double stack   Two etalon filters placed in the same optical path. This arrangement gives a significant reduction in the effective bandwidth.

Dynamo   Mechanism which converts electrical energy into magnetic energy.

Effective temperature   The solar temperature based on its Black Body Curve. Approximately 5780 °K.

Electromagnetic Spectrum   The range of radiation energy which spans from the very short wavelengths (X-rays) to the longer wavelengths of radio. The visible spectrum is a small part of this range.

Ellerman bomb   Bright flash seen at Hα wavelengths associated with the collapse of a magnetic tube.

Emission lines   A bright line seen in a spectrum. Caused by the electrons of various atoms dropping energy levels.

Energy rejection filter   Also called ERF. A filter placed in the optical path of a solar telescope to reduce the incident energy to safe levels.

Entrance Slit   A narrow opening (can be adjustable) positioned at the entrance of a spectroscope to improve the resolution.

Equatorial mount   A telescope mounting which allows tracking of the stars. Usually having two axis of rotation; a Declination axis (Dec) which gives movements North–South and a RA axis which points towards the celestial pole and rotates for tracking.

Etalon   Proper name Fabry–Perot etalon filter. A precision filter made from two reflective glass plates held at precise spacing. Provides a series of very narrowband transmission bands across the spectrum. Usually used in conjunction with a blocking filter.

Evershed flow   Flow of material observed around sunspots.

Exit pupil   The image of the objective seen in an eyepiece. Equal to the size of the objective divided by the magnification of the system.

Facula   Brighter photospheric surface areas associated with sunspots.

Filaments   Long tenuous features seen in the chromosphere and formed near magnetic inversion lines. A cloud like mass of cooler and denser material in the upper chromosphere and corona. They appear as prominences when visible at the solar limb, or filaproms when visible across the limb.

Filaprom   See Filament.

Fibrils   Small dark elongated features seen in Hα wavelengths in and around active areas and sunspots. They are thought to follow the local magnetic field.

Field curvature   An optical aberration. The plane of the best focus position varies from the center to the edge.

Filigree   Unresolved granulation structure visible in the photosphere giving rise to the photospheric network.

First order spectrum   The brightest and least dispersed spectrum formed by a grating.

FITS   Flexible Image Transport system. A digital file structure which allows the maximum information to be recorded. Details of the file contents are encoded into a readable header. Widely used in astronomy.

Flare   Explosive release of energy causing shock waves and the ejection of material. Associated with active areas and can have a lifespan of many hours. Massive flares can give rise to CME and magnetic storms.

Flash spectrum   The emission spectrum from the chromosphere seen in the final moments at the beginning or ending of a total eclipse.

Flat   An image taken of a uniform illuminated surface. Used in imaging to correct vignetting and dust defects.

Flocculi   Large indistinct features in the chromosphere. First noted in the CaK wavelength by G.E. Hale.

Focal length   The distance between the objective lens/mirror of a telescope and the point of focus.

Focal ratio   The focal length of a lens/mirror divided by the aperture.

Focal reducer   A secondary lens which, when placed in a telescope, reduces the effective focal length.

Focus shift   Change in focus position required to bring differing wavelengths into focus. Normally associated with field curvature.

Fraunhofer lines   Prominent dark lines seen in the solar spectrum. First recorded by Joseph Fraunhofer.

Free spectral range   The span of a spectrum which shows no overlap with a subsequent spectral order. Associated with multiple orders from a grating.

Fringing   Dark and light irregular bands seen in some CCD images. Caused by interference patterns generated within the CCD structure when using monochromatic light.

FWHM   Full width half maximum. A measure of the effective size of a spectral line or star image. A measure of the resolution of an optical system.

Gain   Amplification applied to the signal within a CCD. A gain of ×5 will give an ADU readout five times higher than a gain of ×1.

Granulation   Small cellular pattern seen in white light. The individual granules are usually less than 1000 km in size and have a lifetime of a few minutes.

Grating—transmission   A series of grooves (lines) embossed on a plastic layer bonded to a glass support plate. Light is dispersed when it travels through the grating.

Grism   The name given to a small prism when it is combined with a grating to correct the deviation angle.

Heliographic coordinates   The latitude and longitude coordinates used to define the position of a feature on the Sun.

Imaging software   Computer software used to manipulate images. Can also be used to control the camera settings and exposures. i.e. AstroArt5, Maxim DL, IRIS etc.

Instrument response   A spectral profile curve which represents the efficiency of an optical system across the spectral range.

Ion   An atom with one or more electrons removed from the outer shell(s)

Irradiance   The amount of solar energy received by the Earth (W/m^2).

Induced transmission filter   Also ITF. A multi coated filter which only allows a restrict selection of wavelengths to be transmitted.

Jacquinot spot   Also called "sweet spot". A region of uniform illumination seen in an etalon filter field of view.

°Kelvin   Temperature scale based on absolute zero (−273 °C).

Kirchhoff's Laws   Three laws developed to explain the spectral continuum, absorption and emission lines.

Landé factor   A measure of the electron spin and angular momentum. Atoms with high Landé factors make good targets for the measurement of the magnetic Zeeman effect.

Limb   Visible edge of the solar disk.

Linear dispersion   See Plate scale.

Line broadening   The increase in width of a spectral line due to pressure or temperature.

Magnetic storm   Global disruption of the Earth's magnetic field caused by the solar wind. Usually associated with flares and CME.

Magnetograph   A adaptation of the spectroheliograph which measures the solar magnetic field using the Zeeman effect.

Magnetogram   A image showing the magnetic field strength and direction.

Maunder minimum   A period between 1645 and 1715 where there were no visible sunspots or other signs of solar activity.

Micron   Symbol µm, 10^{-6} meter. A common measure for CCD pixel sizes and other small dimensions.

Mono CCD   A CCD where each pixel is responsive to all wavelengths.

Morton wave   The expanding boundary of a shock wave (caused by a magnetic disturbance). Recognized by the Doppler effect.

Mottles   Small scale bright/dark patches associated with spicules. Visible throughout the chromosphere.

Nanometre   Measure of wavelength. 1 nm = 10^{-9} m, = 10 Å.

Network   A cellular structure associated with granular cells and the magnetic field. Visible in both Hα and CaK wavelengths.

Newton Rings   Regular light and dark diffuse lines caused by interference of monochromatic light. Usually between the glass cover plate and the CCD surface. Can affect narrowband imaging in Hα wavelength.

Non uniform spectra   Normally associated with prisms. The spectrum produced is non-linear in scale. The red section is less spread out than the blue section.

Nyquist sampling   Information transfer theory developed by Harry Nyquist. As applied to spectroscopy, it can be used to evaluate the accuracy of the spectrum image. Normally a minimum of 2 pixels are required to determine resolution. Less than two is called "under sampling".

Objective   The main (front) lens (or mirror) of a telescope or camera lens. The size of the objective determines the amount of light collected. Larger objectives allow fainter objects to be seen/recorded as well as provide increased resolution.

Optical aberrations   Image distortions caused by the optical system. These can be chromatic aberrations, coma and field curvature etc.

Optical axis   The virtual line through the center of all the optical components i.e. in a telescope/lens etc. Optical aberrations are usually at a minimum on the optical axis. I.e. coma.

OSC   One shot color. Applied to a CCD which is fitted with a Bayer filter to record images in color.

Penumbra   The outer region of a large sunspot, slightly brighter than the umbral central area. Fine fibrils grains and filaments are usually seen sitting radial across the penumbra.

Photosphere   The white light surface of the Sun where most of the solar energy is released.

Pixel size   An important measure for CCD chips. Measured in micron. I.e. 6.4×6.4 μm pixel. Larger Pixels are usually more sensitive to incoming light. The size of the pixel can affect the final resolution of the image.

Plage   A bright relatively featureless area visible in Hα and CaK wavelengths. Associated with active areas and sunspots.

Planck constant   Symbol $h = 6.6256 \times 10^{-34}$ Js (Joule second). Used in the energy equation, $\Delta E = h\nu$ which relates the energy change (ΔE) of an electron to the frequency (ν) of the emitted/absorbed wavelength of light.

Planck curves   A series of spectrum intensity curves which show the peak wavelength position changes with temperature.

Plasma   Ionized gas usually electrons and positive ions.

Plate Scale   Also called linear dispersion. The scale, in Å/mm or Å/pixel of a spectrum. The higher the plate scale, the smaller the size of the spectrum and the lower the final resolution.

Point spread function   The distribution of intensity resulting from the use of multiple optical elements or distortions within the light path (due to seeing conditions).

Pore   Small dark mark visible in the photosphere. Usually the precursor for sunspot growth.

Prism   Shaped piece of glass which produces a spectrum due to dispersion. Can be used with a grating to compensate for the deviation angle (Grism).

Prominence   See Filament.

Quantum Theory   A branch of physics which deals with small particles. Uses both wave and particle analogies to resolve interactions at a sub-atomic level.

Radiative zone   The inner 70 % of the solar core where energy is transferred through radiation.

RAW A proprietary image format used by manufacturers of DSLR cameras. It contains additional information on each color channel and provides the maximum amount of image data.

R value A measure of the resolving power of a spectrograph. Equal to the wavelength (λ) of a pair of lines, just separated, divided by the distance between them ($\Delta\lambda$). I.e. $R = \lambda/\Delta\lambda$. $R = 100$ is a low resolution spectroscope. $R = 40000$ is a very high resolution instrument.

Reflecting telescope A telescope which uses mirrors to form the image. I.e. Newtonian. Does not suffer from chromatic aberration, all wavelengths of light come to a common focus.

Refractive index Symbol n, A measure of the dispersing power of a prism. The refractive index varies with the wavelength of light. Glass has a typical index of $n = 1.5$

Resolution The ability to see two close lines as separate lines. Measured by the smallest distance between them, normally in Å

Reversal layer Name given to the lower region of the chromosphere by early observers. Where the flash spectrum occurred.

RGB Red, Green, Blue. The sequence of color filters used in the Bayer Matrix.

SCT Schmidt Cassegrain Telescope. A corrector lens is used with a spherical mirror to provide an optically corrected image. Usual focal ratio of f10. I.e. Meade/Celestron.

Seeing The size of a star's disk as seen through a telescope. The size varies with atmospheric conditions. Usually measured in arc seconds.

Sidereal tracking An astronomical mount which is fitted with a drive motor. This rotates the polar axis in RA at a rate of 360° per sidereal day (23 h 56 m 4 s) thereby keeping track with the celestial movement of the sky.

SNR Signal to noise ratio. A measure of the unwanted interference (noise) found in images and spectra. A SNR > 50 is normally required in results for professional acceptance.

Solar cycle A cyclic period of solar activity (evidenced by the number of sunspots) of 11 years.

Solar wind A outward flow of electrons and protons from the sun carrying the solar magnetic field well beyond the solar system.

Spectral coma An optical aberration where the edge of the field of view shows distorted images which vary in size with color.

Spectral lines Bright or dark lines seen in a spectrum. Caused by the absorption (or emission) of energy by atoms.

Spectrograph Instrument containing a dispersion element (prism or grating) to form an image of the spectrum.

Spectroheliograph Also SHG. A combination of telescope spectrograph and imaging camera which allows the recording of the solar features in a limited bandwidth (Usually Hα and CaK).

Spectrohelioscope Also SHS. Adaptation of the spectroheliograph to include a synthesizer arrangement which allows the direct observation (generally in Hα) of solar features due to persistence of vision.

Spicule Small spike like feature which generally cover the whole of the Hα chromospheric surface. Can sometimes be recorded in pairs and clumps. Can extend 1000 km upwards and have lifetimes of a few minutes.

Spray See Filament.

Star classification A sequence of star spectra based on absorption features. Originally based on the visibility and clarity of hydrogen absorption lines. The classification sequence from hot to cooler stars is OBAFGKM. The Sun is a G2V star.

Stoneyhurst disk A series of charts prepared for a year showing a the heliographic lines of latitude and longitude.

Sunspot A darker cooler area in the photosphere caused by the solar magnetic field. Also visible as dark disruptions in Hα and Cak wavelengths.

Supergiant Large luminous stars with absolute magnitudes greater than −4.6. The majority of supergiants are K and M type stars. They are identified in the star classification system with the Suffixes Ia and IIa i.e. M5 Ia

Supergranulation A convection pattern of large (30,000 km) cells visible in white light.

Telluric lines Absorption lines in the spectrum caused by oxygen and water vapor in the Earth's atmosphere.

Tiff Tagged Image file format. Uncompressed high quality image format.

Tower telescope A popular solar telescope design in the twentieth century. The height of the tower and the vertical arrangement of the optics were designed to minimize thermal (seeing) effects. First built by G.E. Hale at Mt. Wilson in 1907.

Transition region The upper region of the chromosphere where there is a rapid temperature increase towards the Corona (from 20,000 °K to greater than 2 million °K).

Transversalium Dark lines see running the length of a spectral image. Caused by dust or inaccuracies in the edges of the entrance slit of a spectrograph.

T thread A M42×0.75 mm pitch thread, commonly used on camera adaptors.

Umbra The dark core area of a sunspot. Can sometimes be seen crossed by filaments or mini flares "light bridges". A cooler region (4100 °K) with strong magnetic fields.

Vignetting Light loss in an optical system caused by interference to the light path. Usually seen as darker edges in an image.

Visible spectrum The part of the electromagnetic spectrum visible to the human eye. Usually from 4000 to 7000 Å. Also called the optical spectrum.

Wavelength A characteristic of electromagnetic energy. Short wave lengths give gamma rays, X-rays etc. and long wavelengths Infrared and radio. A small portion of the electromagnetic energies with wavelengths between 4000 and 7000 Å are visible (optical or visual spectrum).

Wave number Symbol $\tilde{v}$ ($\tilde{v}= 1/\lambda$ cm^{-1}), the number of wavelengths per unit distance (the number of cycles per wavelength), where λ is the wavelength.

White light Visible wavelengths between 4000 and 7000 Å.

Wilson effect The displacement of the umbra towards the center of the solar disk. Possibly caused by depression of the umbra area.

Zeeman effect The separation of absorption lines due to magnetic fields.

Zero order image The undeviated image of the entrance slit seen with a grating. The zero order image lies on the optical axis and represents a zero wavelength position. i.e. for calibration purposes, zero order wavelength, $\lambda_0 = 0.00$ Å.

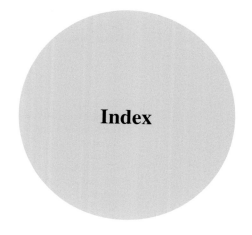

Index

© Springer International Publishing Switzerland 2016
K.M. Harrison, *Imaging Sunlight Using a Digital Spectroheliograph*,
The Patrick Moore Practical Astronomy Series, DOI 10.1007/978-3-319-24874-5

Printed in the United States
By Bookmasters